5th Social Media Mining for Health Applications Workshop & Shared Task 2020

Held online due to COVID-19

Barcelona, Spain
12 December 2020

ISBN: 978-1-7138-2820-4

COLING 2020

Social Media Mining for Health Applications (#SMM4H) Workshop & Shared Task

Proceedings of the Fifth Workshop

December 12, 2020
Barcelona, Spain (Online)

Preface

Welcome to the 5th Social Media Mining for Health Applications Workshop & Shared Task (#SMM4H 2020), co-located at the 28th International Conference on Computational Linguistics (COLING 2020). In its fifth iteration, #SMM4H 2020 continues to serve as a venue for bringing together researchers interested in addressing the significant opportunities and challenges of utilizing the vast amount of data on social media for health informatics. For #SMM4H 2020, we accepted 5 workshop papers (acceptance rate of 56%) and 26 shared task system description papers. Each submission was peer-reviewed by two reviewers.

The accepted workshop papers span a range of social media data—Twitter, Facebook, Reddit, and online health forums—and health domains, including diabetes, depression, COVID-19, medical misinformation, and adverse drug reactions. Cornelius et al. present an online platform that aggregates and visualizes methods for extracting information related to COVID-19 on Twitter. Dirkson et al. explore modeling conversational features of posts, in addition to the posts themselves, for detecting adverse drug reactions on Facebook, and medical misinformation. Romberg et al. present an annotated, German-language corpus for extracting information needs expressed online by patients with diabetes. Moßburger et al. use various text mining techniques to compare features of depression forums on Reddit and a curated, moderated site. Finally, Owen et al. present an annotated, English-language corpus for detecting depression and anxiety on Twitter.

The #SMM4H 2020 shared tasks sought to advance the use of Twitter data (tweets) for pharmacovigilance, toxicovigilance, and epidemiology of birth defects. In addition to re-reruns of three tasks, #SMM4H 2020 included new tasks for detecting adverse drug reactions in French and Russian tweets, characterizing chatter related to prescription medication abuse, and detecting self reports of birth defect pregnancy outcomes. The five tasks required methods for binary classification, multi-class classification, and named entity recognition (NER). With 29 teams and a total of 130 system submissions, participation in the #SMM4H shared tasks continues to grow. Among the 26 shared task system description papers that were accepted, 6 teams were invited to present their system orally.

The organizing committee of #SMM4H 2020 would like to thank the program committee, the additional reviewers of system description papers, the organizers of COLING 2020 (especially the workshop co-chairs), the annotators of the shared task data, and, of course, everyone who submitted a paper or participated in the shared tasks. #SMM4H 2020 would not have been possible without all of them.

Graciela, Ari, Ivan, Davy, Arjun, Karen, Abeed, Anne-Lyse, Elena, Zulfat, Ilseyar

Organizing and Program Committees

Organizing Committee:

Graciela Gonzalez-Hernandez, University of Pennsylvania, USA
Ari Z. Klein, University of Pennsylvania, USA
Ivan Flores, University of Pennsylvania, USA
Davy Weissenbacher, University of Pennsylvania, USA
Arjun Magge, University of Pennsylvania, USA
Karen O'Connor, University of Pennsylvania, USA
Abeed Sarker, Emory University, USA
Anne-Lyse Minard, Université d'Orléans, France
Elena Tutubalina, Kazan Federal University, Russia
Zulfat Miftahutdinov, Kazan Federal University, Russia
Ilseyar Alimova, Kazan Federal University, Russia

Program Committee:

Robert Leaman, US National Library of Medicine, USA
Diego Molla, Macquarie University, Australia
Azadeh Nikfarjam, Apple, USA
Thierry Poibeau, French National Center for Scientific Research, France
Kirk Roberts, University of Texas Health Science Center at Houston, USA
Yutaka Sasaki, Toyota Technological Institute, Japan
H. Andrew Schwartz, Stony Brook University, USA
Nicolas Turenne, French National Institute for Agricultural Research, France
Karin Verspoor, University of Melbourne, Australia
Pierre Zweigenbaum, French National Center for Scientific Research, France

Table of Contents

Workshop Program

Saturday, December 12, 2020

14:00–14:10 *Introduction*
Graciela Gonzalez-Hernandez

14:10–14:50 **Oral Presentations Q&A Session 1**

COVID-19 Twitter Monitor: Aggregating and Visualizing COVID-19 Related Trends in Social Media
Joseph Cornelius, Tilia Ellendorff, Lenz Furrer and Fabio Rinaldi

Conversation-Aware Filtering of Online Patient Forum Messages
Anne Dirkson, Suzan Verberne and Wessel Kraaij

Annotating Patient Information Needs in Online Diabetes Forums
Julia Romberg, Jan Dyczmons, Sandra Olivia Borgmann, Jana Sommer, Markus Vomhof, Cecilia Brunoni, Ismael Bruck-Ramisch, Luis Enders, Andrea Icks and Stefan Conrad

Overview of the Fifth Social Media Mining for Health Applications (#SMM4H) Shared Tasks at COLING 2020
Ari Klein, Ilseyar Alimova, Ivan Flores, Arjun Magge, Zulfat Miftahutdinov, Anne-Lyse Minard, Karen O'Connor, Abeed Sarker, Elena Tutubalina, Davy Weissenbacher and Graciela Gonzalez-Hernandez

14:50–15:00 **Break**

15:00–15:40 *Invited Talk*
Fabio Rinaldi

15:40–15:50 **Break**

15:50–16:30 **Oral Presentations Q&A Session 2**

Ensemble BERT for Classifying Medication-mentioning Tweets
Huong Dang, Kahyun Lee, Sam Henry and Özlem Uzuner

ISLab System for SMM4H Shared Task 2020
Chen-Kai Wang, Hong-Jie Dai, You-Chen Zhang, Bo-Chun Xu, Bo-Hong Wang, You-Ning Xu, Po-Hao Chen and Chung-Hong Lee

BERT Implementation for Detecting Adverse Drug Effects Mentions in Russian
Andrey Gusev, Anna Kuznetsova, Anna Polyanskaya and Egor Yatsishin

KFU NLP Team at SMM4H 2020 Tasks: Cross-lingual Transfer Learning with Pre-trained Language Models for Drug Reactions
Zulfat Miftahutdinov, Andrey Sakhovskiy and Elena Tutubalina

16:30–17:30   **Poster Session**

17:30–17:40   **Break**

17:40–18:20   **Oral Presentations Q&A Session 3**

SMM4H Shared Task 2020 - A Hybrid Pipeline for Identifying Prescription Drug Abuse from Twitter: Machine Learning, Deep Learning, and Post-Processing
Isabel Metzger, Emir Y. Haskovic, Allison Black, Whitley M. Yi, Rajat S. Chandra, Mark T. Rutledge, William McMahon and Yindalon Aphinyanaphongs

Automatic Detecting for Health-related Twitter Data with BioBERT
Yang Bai and Xiaobing Zhou

Exploring Online Depression Forums via Text Mining: A Comparison of Reddit and a Curated Online Forum
Luis Moßburger, Felix Wende, Kay Brinkmann and Thomas Schmidt

Towards Preemptive Detection of Depression and Anxiety in Twitter
David Owen, Jose Camacho-Collados and Luis Espinosa Anke

18:20–18:30   *Conclusion*
Graciela Gonzalez-Hernandez

COVID-19 Twitter Monitor: Aggregating and visualizing COVID-19 related trends in social media

Joseph Cornelius[*]   **Tilia Ellendorff**[†‡]   **Lenz Furrer**[†‡]   **Fabio Rinaldi**[*†‡]

[*]Dalle Molle Institute for Artificial Intelligence Research (IDSIA)
[†]Swiss Institute of Bioinformatics
[‡]University of Zurich, Department of Computational Linguistics
{joseph.cornelius,fabio.rinaldi}@idsia.ch
{tilia.ellendorff, lenz.furrer}@uzh.ch

Abstract

Social media platforms offer extensive information about the development of the COVID-19 pandemic and the current state of public health. In recent years, the Natural Language Processing community has developed a variety of methods to extract health-related information from posts on social media platforms. In order for these techniques to be used by a broad public, they must be aggregated and presented in a user-friendly way. We have aggregated ten methods to analyze tweets related to the COVID-19 pandemic, and present interactive visualizations of the results on our online platform, the COVID-19 Twitter Monitor. In the current version of our platform, we offer distinct methods for the inspection of the dataset, at different levels: corpus-wide, single post, and spans within each post. Besides, we allow the combination of different methods to enable a more selective acquisition of knowledge. Through the visual and interactive combination of various methods, interconnections in the different outputs can be revealed.

1 Introduction

Today, social media platforms are important sources of information, with a usage rate of over 70% for adults in the USA and a continuous increase in popularity.[1] Platforms like Twitter are characterized by their thematic diversity and realtime coverage of worldwide events. As a result, social media platforms offer information, especially about ongoing events that are locally and dynamically fluctuating, such as the SARS-CoV-2 (COVID-19) pandemic. The micro-posts contain not only the latest news or announcements of new medical findings but also reports about personal well-being and the sentiment towards public health interventions. Information contained in tweets has multiple usages. For instance, domain experts could discover how recent scientific studies find an echo in social media, while the health industry could find patients' reactions to certain drugs, and the general public could observe trends towards popular topics (e.g. the alternating popularity of politicians over time).

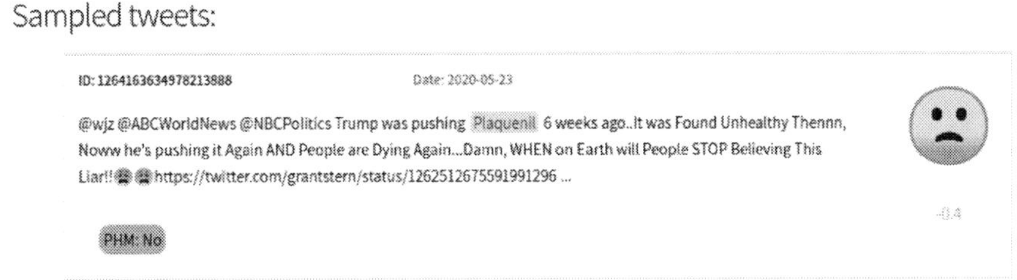

Figure 1: A single COVID-19 related tweet annotated with our drug brand name detection system and classified by our systems for personal health mention (PHM) identification and sentiment analysis.

[1]https://www.pewresearch.org/internet/fact-sheet/social-media/

Proceedings of the 5th Social Media Mining for Health Applications (#SMM4H) Workshop & Shared Task, pages 1–10
Barcelona, Spain (Online), December 12, 2020.

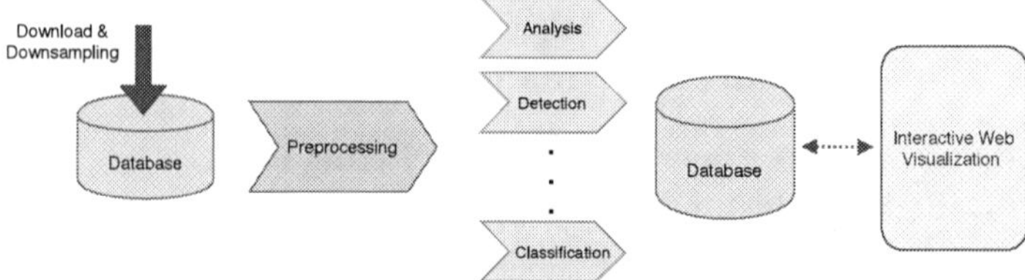

Figure 2: A schematic of the data acquisition and processing pipeline of our web platform.

To provide universal access to the acquisition of knowledge from public discourse about COVID-19 on Twitter, social media mining methods have to be easily applicable.

We present a platform called COVID-19 Twitter Monitor that analyzes, categorizes, and extracts textual information from posts on Twitter. Our platform provides insights into the development and spread of the COVID-19 pandemic. The information is retrieved at different levels: spans within each post (e.g. drug name detection), single post (e.g. text language identification) and corpus level (e.g. distribution of hashtags). We use a variety of different methods to analyse pandemic-related micro-posts with a focus on health issues.

In contrast to existing platforms, we aim to provide a platform for newly emerging diseases for which no disease-specific supervised datasets are available. Therefore, our approach concentrates on unsupervised and pre-trained supervised methods that can be expected to generalize to unseen conditions. We have created an interactive web interface that integrates these methods and thus simplifies the use of social media mining methods. When visualizing the results, different methods are combined to display underlying connections in the extracted information. With this approach, we hope to make the contained information more accessible and provide an increased level of insight. Since we present the first version of our web platform[2], this is only an initial description of our approach and the preliminary results.

2 Related Work

Numerous studies show that social media platforms are particularly suitable for obtaining health-related information like the identification of drug abuse or the tracking of infectious diseases (Woo et al., 2016; Paul et al., 2016; McGough et al., 2017; Cocos et al., 2017). The detection of infectious disease outbreaks by using data from Twitter has already been tested, for instance in the case of the Ebola virus (Odlum and Yoon, 2015) and the Zika virus (Juric et al., 2017; Mamidi et al., 2019). Paul et al. (2016) propose using social media data for disease surveillance by locally determining and forecasting influenza prevalence. They address obstacles that the typically informal text on social media platforms presents to common language processing systems, and stress the need for normalized medical terminology and general linguistic normalization for social media. In addition, they argue that sentiment analysis is particularly suitable for analyzing the patient's impression of the quality of health care or public opinion on certain drugs. Likewise, Mamidi et al. (2019) apply sentiment analysis to tweets concerning virus outbreaks to gain information for disease control.

Joshi et al. (2019) describe how to identify disease outbreaks using text-based data. Like Wang et al. (2014), they show how topic modeling can be used to discover health-related topics in the micro-posts. Twitter adapted topic models were also utilized by Paul and Dredze (2011) to conduct syndromic surveillance regarding the flu.

Lamb et al. (2013) showed that more intricate data retrieval techniques can be applied to noisy Twitter data, such as identifying posts that mention personal health (PHM). Further fine-grained analyses were conducted by them, for example, to determine whether the tweet concerns the health of the author or a person familiar with the author. Going even further, Yin et al. (2015) showed that Twitter-related PHM classifiers can operate disease agnostic; they trained their system on four disease contexts and could achieve a precision of 0.77 on 34 other health issues.

Several platforms have already explored using social media data for health surveillance and disease

---

[2]https://covid19smm.nlp.idsia.ch

	total number	unique	tweets containing
URLs	372228	283887	323760 *(72.76%)*
Domains	372228	37282	323760 *(72.76%)*
Hashtags	380744	93454	136279 *(30.63%)*
Languages	425721	53	425721 *(95.67%)*
RxNorm Brand Names	1285	101	1218 *(0.27%)*
Preprint Paper	152	124	149 *(0.03%)*
Tweets	444990	444990	–

Table 1: Statistics of the data collection with formerly 500K tweets after similarity filtering. For each feature, the total number of occurrences, the number of distinct occurrences, and the total number of tweets containing the feature are displayed.

tracking. Lee et al. (2013) uses Twitter data for influenza and cancer surveillance but focuses on the computation of various distributions, e.g. the temporal change in quantity or the most frequent words of influenza tweets. In contrast, HealthMap[3] (Brownstein et al., 2008) lets you explore various disease outbreaks at the corpus and post level, but primarily uses google news as a data source. Systems such as Flutrack (Chorianopoulos and Talvis, 2016) and Influenza Observations and Forecast[4] focus instead on map-based identification of infection disease outbreaks.

3   Data

We use the COVID-19 Twitter dataset published by the Panacea Lab (Banda et al., 2020). The dataset consists of daily Twitter chatter (about 4M tweets per day), collected by the Twitter Stream API for the keywords "COVD19", "CoronavirusPandemic", "COVID-19", "2019nCOV", "CoronaOutbreak", "coronavirus", and "WuhanVirus". In addition, the Panacea Lab extended the dataset over time with Twitter datasets of research groups from the Universitat Autònoma de Barcelona, the National Research University Higher School of Economics, and the Kazan Federal University. As of June 14, 2020, version 14.0 of the dataset contains over 400 million unique tweets. Since the data has been collected irrespective of language, all languages are covered, with a pronounced prevalence of English (60%), Spanish (16%), and French (4%) tweets. An update is provided every two days and a cumulative update is provided every week. We use the cleaned version of the dataset provided by the Panacea Lab without retweets.

4   Methods and Results

In the following section we describe the different methods used by our platform. We start by creating a randomly selected subset of the Panacea Lab's COVID-19 Twitter dataset.[5] We then process this subset with the preprocessing steps described below and apply the methods for the data analysis as described in this section. Subsequently, we store the obtained results and the original data subset in a database accessed by our interactive web platform for visualization of the results, as illustrated in Figure 2.

4.1   Pre-Processing

For tweets used for topic modeling, sentiment analysis, and PHM identification, our system applies the following preprocessing methods:

- Without splitting the sentences, all tweets are tokenized using the spaCy[6] tokenizer.
- URLs are reduced to their domain names.
- The hash symbol "#" is removed from all hashtags.
- Multiple whitespace characters are removed.

---

[3] https://healthmap.org
[4] http://cpid.iri.columbia.edu
[5] This step is made necessary by the limitations of the computational infrastructure at our disposal for this activity.
[6] https://spacy.io/api/tokenizer

- Numbers are substituted with the token NUMBER.
- All @username are standardized to @USER.
- Camel-cased tokens are split into their components, e.g. "VirusOutbreak" to "Virus Outbreak".
- Colloquial abbreviations such as "w/" for "with" are resolved.

4.2 Collection statistics

To reveal the following aspects in the dataset, we utilize distributions of multiple features. To identify different trending (sub)topics, we apply hashtag distributions and the domain/URL distribution to discover frequently-used sources and references of the tweet content. Furthermore, we use the distribution of languages in the dataset as a proxy to estimate the prevalence of the pandemic in different cultures and countries. To show the history of the intensity of discussion on different subtopics, we utilize the time distribution of tweets.

4.2.1 Hashtag Distribution

We extract all hashtags from each tweet in the corpus for the calculation of the hashtag distribution. In order to normalize the hashtags, we separate the hash symbol ("#") from the hashtags, split all the hashtags written in camel-case into the individual parts and lowercase them. We utilize the resulting hashtags to compute the hashtag distribution. As shown in Table 1, with over 93K distinct hashtags in our subset with around 445K tweets, the data collection contains a broad diversity of subtopics.

4.2.2 Domain and URL Distribution

To detect URLs in tweets, we filter all posts for tokens that start with "http://", "https://", or "www.". In addition, we use an unshortening method for each link, i.e. we retrieve the redirect information to get the target URL of links which have been compressed by the link shortening service (e.g. bitly.com). We derive the URL distribution from the processed URLs. By additionally reducing the URLs to their top-level domain (TLD) we get the domain distribution.

4.2.3 Language Distribution

For language detection, we use Nakatani Shuyo's port to Python of the Language Detection Library for Java (Nakatani, 2010). The language detection is based on naive Bayesian filtering and is trained on

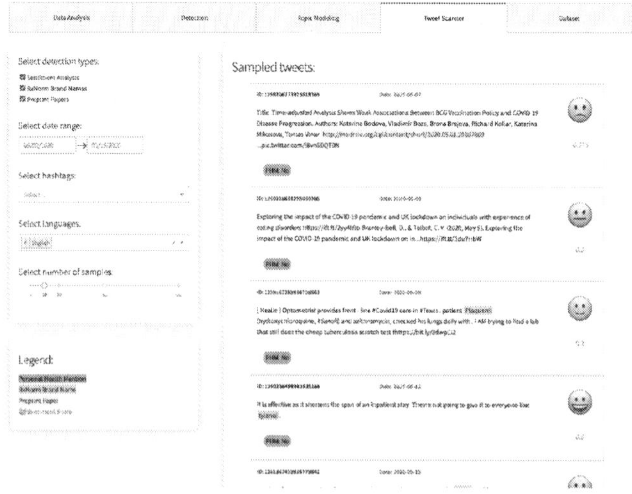

Figure 3: The screenshot of our web platform shows the subpage for the analysis of individual tweets. To obtain a selection of tweets, we filter them according to the analysis method, the date of creation, the contained hashtags, and their language. The selected tweets and the graphical indication of the different analysis results are displayed.

language profiles produced by Wikipedia abstracts. The classifier achieves an accuracy of 99% for 53 different languages. We produce the language distribution from the resulting language classification of each tweet. Our data collection contains tweets from all 53 languages, as can be seen in Table 1.

4.2.4 Temporal Tweet Distribution

We grouped the tweets by day and hour to display the temporal distribution and the temporal progression of tweets. Besides, we also allow the selection for tweets of a specific hashtag, see in Figure 5.

4.3 Topic Modeling

One of the hallmarks of Twitter is that an extensive discourse often accompanies emerging events. One method to find different themes within a vast amount of unlabeled posts is topic modeling. The goal of topic modeling is to discover in a probabilistic fashion distinct thematic structures in the underlying text corpus. It allows us to identify the main underlying topics that the collection of tweets is concerned with. We use the probabilistic, generative, and unsupervised Latent Dirichlet allocation (LDA) as a topic model (Blei et al., 2003; Hoffman et al., 2010).

We assume that each tweet contains a mixture of different topics in different proportions. The topics are considered to be a combination of different words. Since the number of topics is not known, we set the number of topics to twelve, a value we obtained by manual approximation. We utilize the LDA method of the python scikit-learn library.[7]

By means of two example topics found by the LDA method ["china", "coronavirus", "wuhan", "spread", "chinese", "outbreak",...] and ["mask", "coronavirus", "face", "vaccine", "reopen", "wear",...] it is possible to see how we can identify different thematic discussions within the tweets' collection.

4.4 Drug Brand Name Detection

For the detection of drug brand names, we use the OntoGene Entity Recognizer (OGER) (Furrer and Rinaldi, 2017; Basaldella et al., 2017; Furrer et al., 2019). OGER is tailored for biomedical entity recognition and uses a combination of dictionary-based lookup and flexible matching. As dictionary, we use the RxNorm terminology,[8] a US-specific terminology that provides normalized names for all clinical drugs available on the US market. In the dictionary-based lookup for drug names, we consider only entries with the term types corresponding to brand names, since this allows us to significantly reduce the number of mismatches with a minor reduction of matches. In addition, we use a manually constructed blacklist with drug brand names that are ambiguous, such as "Android", which is a mobile operating system as well as a brand name under which "methyltestosterone" is sold. To determine the precision of brand name detection, we manually evaluated 400 tweets containing tokens classified by our system as a brand name. The system achieves a precision of 0.67; however, 301 tweets are removed by prior filtering with the blacklist.[9]

4.5 Sentiment Analysis

Sentiment analysis is the interpretation and classification of emotions, used to differentiate the tweets into positive, negative, and neutral. We score the sentiment of each tweet in the range of [-1,+1] with -1 negative, 0 neutral, and +1 positive, by employing the Transformer based sentiment analysis model by Flair (Akbik et al., 2018). This model is trained on the IMDB review dataset (Maas et al., 2011) and achieves an accuracy of 98.87%. In addition to the sentiment analysis of individual tweets, we have an interactive panel to visualize the temporal variation of the sentiment score for tweets containing a selected hashtag. Furthermore, we calculate the average sentiment score over the entire period. This allows the user to track the sentiment of all tweets with specific hashtags, enabling the monitoring of changes in sentiment related to various topics in the public health discourse. For example, #mentalhealth

---

[7] https://scikit-learn.org/stable/modules/generated/sklearn.decomposition.LatentDirichletAllocation.html

[8] https://www.nlm.nih.gov/research/umls/rxnorm/index.html

[9] The following words are included in the blacklist: "android", "schiff", "fml", "pronto", "alli", "icar", "cambia", "ssd", "tabloid", "vanquish", "ting", and "propel".

emerged on March 16th with a strongly negative sentiment four days after the first tweet with #lockdown occurred with slightly more positive sentiment in the data collection, one day after the WHO declared COVID-19 a pandemic.

4.6 Preprint Paper Detection

For the detection of URLs referring to preprint papers, we use the unshortened URLs of the tweets and extract the top-level domains (TLDs). We then compare them against a manually curated list of 55 TLDs of preprint servers. We can estimate the popularity of a preprint paper by the frequency of its mention in different tweets. Therefore we measure the string similarity[10] between all URLs to count and cluster URLs link to the same preprint paper. Our system reveals over 150 URLs of preprint paper in the subset of approx. 445K tweets, see Table 1.

4.7 Personal Health Mention Identification

In contrast to tweets that address general awareness of health issues, tweets about PHMs are concerned with the author's own health issues or those of persons familiar to the author. We use a BERT-based model for the identification of PHMs (Ellendorff et al., 2019). The model is trained to generalize to unseen health contexts. It was trained on two Twitter datasets belonging to two different flu-related contexts, flu infection and flu vaccination. The model was developed within the SMM4H shared task in 2019 and scored the best results. It achieved an overall accuracy of 0.877 and an F-score of 0.873 on the official test set, which includes tweets from seen and unseen health contexts.

4.8 Visualization

For the visualization of our interactive web interface, we use Dash[11]. We have currently limited the size of the dataset to 500K tweets to allow a prompt processing of the data and a seamless web visualization of the results. For a clearer presentation we display the calculation in five categorized tabs:

- **Data Statistics**: a tab that shows the distributions over the whole dataset, see Figure 6.
- **Detection**: a subpage for all detection methods that operate on the whole dataset (e.g. sentiment analysis, preprint paper identification, and medication brand name detection) as shown in Figure 4.
- **Topic Modeling**: a visualization of the results of the topic modeling using the pyLDAvis library.[12]
- **Tweet Scanner**: a subpage to show the detection methods on the tweet level, see Figure 3 and 1.
- **Dataset**: a panel with more detailed information about the tweets contained in the dataset (e.g. Tweet-ID and date).

5 Conclusion and Future Work

In this paper, we have presented the first version of our interactive web platform for the aggregation and visualization of different methods for social media mining regarding COVID-19.

Concerning social media mining, one of the biggest challenges in the analysis of emerging pandemics such as COVID-19 is that we initially have a very limited number of supervised datasets. Therefore, the selected methods applied to our platform are either unsupervised or have been previously trained on other social media datasets. We have demonstrated that by the combined visualization of different methods, underlying connections in the data can be revealed. In particular, we aimed to make the interactive web platform comprehensible for both a specialist audience and the general public.

As future work we intend to move the system to a more capable platform in order to be able to include a much larger number of tweets, and to add novel capabilities, such as the detection of tweets generated by bots as opposed to those generated by humans, through the usage of a tool such as the Botometer.[13]

---

[10]The string similarity is measured with the SequenceMatcher module `https://docs.python.org/3.6/library/difflib.html`

[11]`https://plotly.com/dash`

[12]`https://github.com/bmabey/pyLDAvis`

[13]`https://botometer.iuni.iu.edu/`

References

Alan Akbik, Duncan Blythe, and Roland Vollgraf. 2018. Contextual string embeddings for sequence labeling. In *COLING 2018, 27th International Conference on Computational Linguistics*, pages 1638–1649.

Juan M. Banda, Ramya Tekumalla, Guanyu Wang, Jingyuan Yu, Tuo Liu, Yuning Ding, and Gerardo Chowell. 2020. A large-scale COVID-19 Twitter chatter dataset for open scientific research – an international collaboration.

Marco Basaldella, Lenz Furrer, Carlo Tasso, and Fabio Rinaldi. 2017. Entity recognition in the biomedical domain using a hybrid approach. *Journal of Biomedical Semantics*, 8(1):51.

David M Blei, Andrew Y Ng, and Michael I Jordan. 2003. Latent Dirichlet allocation. *Journal of Machine Learning Research*, 3(Jan):993–1022.

John S Brownstein, Clark C Freifeld, Ben Y Reis, and Kenneth D Mandl. 2008. Surveillance Sans Frontieres: Internet-based emerging infectious disease intelligence and the HealthMap project. *PLoS Med*, 5(7):e151.

Konstantinos Chorianopoulos and Karolos Talvis. 2016. Flutrack.org: Open-source and linked data for epidemiology. *Health informatics journal*, 22(4):962–974.

Anne Cocos, Alexander G Fiks, and Aaron J Masino. 2017. Deep learning for pharmacovigilance: recurrent neural network architectures for labeling adverse drug reactions in Twitter posts. *Journal of the American Medical Informatics Association*, 24(4):813–821.

Tilia Ellendorff, Lenz Furrer, Nicola Colic, Noëmi Aepli, and Fabio Rinaldi. 2019. Approaching SMM4H with merged models and multi-task learning. In *Proceedings of the Fourth Social Media Mining for Health Applications (#SMM4H) Workshop & Shared Task*, pages 58–61. Association for Computational Linguistics.

Lenz Furrer and Fabio Rinaldi. 2017. OGER: OntoGene's entity recogniser in the BeCalm TIPS task. In *Proceedings of the BioCreative V.5 Challenge Evaluation Workshop*, pages 175–182.

Lenz Furrer, Anna Jancso, Nicola Colic, and Fabio Rinaldi. 2019. OGER++: hybrid multi-type entity recognition. *Journal of Cheminformatics*, 11(1):7.

Matthew Hoffman, Francis R Bach, and David M Blei. 2010. Online learning for latent dirichlet allocation. In *advances in neural information processing systems*, pages 856–864.

Aditya Joshi, Sarvnaz Karimi, Ross Sparks, Cécile Paris, and C Raina MacIntyre. 2019. Survey of Text-based Epidemic Intelligence: A computational linguistics perspective. *ACM Computing Surveys (CSUR)*, 52(6):1–19.

Radmila Juric, Inhwa Kim, Hemalatha Panneerselvam, and Igor Tesanovic. 2017. Analysis of Zika virus tweets: Could hadoop platform help in global health management? In *Proceedings of the 50th Hawaii International Conference on System Sciences*.

Alex Lamb, Michael Paul, and Mark Dredze. 2013. Separating fact from fear: Tracking flu infections on Twitter. In *Proceedings of the 2013 Conference of the North American Chapter of the Association for Computational Linguistics: Human Language Technologies*, pages 789–795.

Kathy Lee, Ankit Agrawal, and Alok Choudhary. 2013. Real-time disease surveillance using Twitter data: demonstration on flu and cancer. In *Proceedings of the 19th ACM SIGKDD international conference on Knowledge discovery and data mining*, pages 1474–1477.

Andrew L. Maas, Raymond E. Daly, Peter T. Pham, Dan Huang, Andrew Y. Ng, and Christopher Potts. 2011. Learning word vectors for sentiment analysis. In *Proceedings of the 49th Annual Meeting of the Association for Computational Linguistics: Human Language Technologies*, pages 142–150. Association for Computational Linguistics.

Ravali Mamidi, Michele Miller, Tanvi Banerjee, William Romine, and Amit Sheth. 2019. Identifying key topics bearing negative sentiment on Twitter: insights concerning the 2015-2016 Zika epidemic. *JMIR Public Health and Surveillance*, 5(2):e11036.

Sarah F McGough, John S Brownstein, Jared B Hawkins, and Mauricio Santillana. 2017. Forecasting Zika incidence in the 2016 Latin America outbreak combining traditional disease surveillance with search, social media, and news report data. *PLoS neglected tropical diseases*, 11(1):e0005295.

Shuyo Nakatani. 2010. Language detection library for Java.

Michelle Odlum and Sunmoo Yoon. 2015. What can we learn about the Ebola outbreak from tweets? *American journal of infection control*, 43(6):563–571.

Michael J Paul and Mark Dredze. 2011. You are what you tweet: Analyzing Twitter for public health. In *Fifth International AAAI Conference on Weblogs and Social Media*.

Michael J Paul, Abeed Sarker, John S Brownstein, Azadeh Nikfarjam, Matthew Scotch, Karen L Smith, and Graciela Gonzalez. 2016. Social media mining for public health monitoring and surveillance. In *Biocomputing 2016: Proceedings of the Pacific symposium*, pages 468–479.

Shiliang Wang, Michael J Paul, and Mark Dredze. 2014. Exploring health topics in chinese social media: An analysis of Sina Weibo. In *Workshops at the Twenty-Eighth AAAI Conference on Artificial Intelligence*.

Hyekyung Woo, Youngtae Cho, Eunyoung Shim, Jong-Koo Lee, Chang-Gun Lee, and Seong Hwan Kim. 2016. Estimating influenza outbreaks using both search engine query data and social media data in South Korea. *Journal of medical Internet research*, 18(7):e177.

Zhijun Yin, Daniel Fabbri, S Trent Rosenbloom, and Bradley Malin. 2015. A scalable framework to detect personal health mentions on Twitter. *Journal of medical Internet research*, 17(6):e138.

A   Appendices

Figure 4: The screenshot of our web platform displays the subpage for the corpus-based detection methods. It shows the temporal sentiment analysis selected by hashtags, the detection of drug brand names, and the detection of preprint papers, respectively.

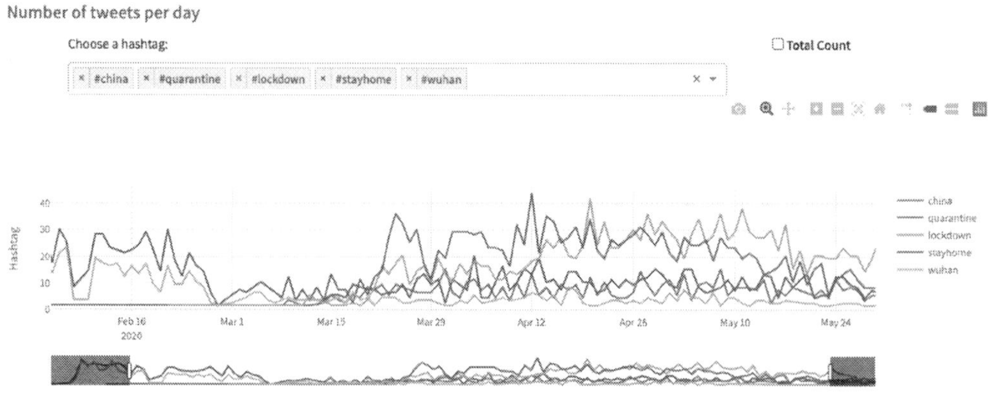

Figure 5: The detail from the screenshot of our web platform displays the temporal distribution of tweets. The progression can be visualized for all tweets as well as for tweets selected by their hashtags.

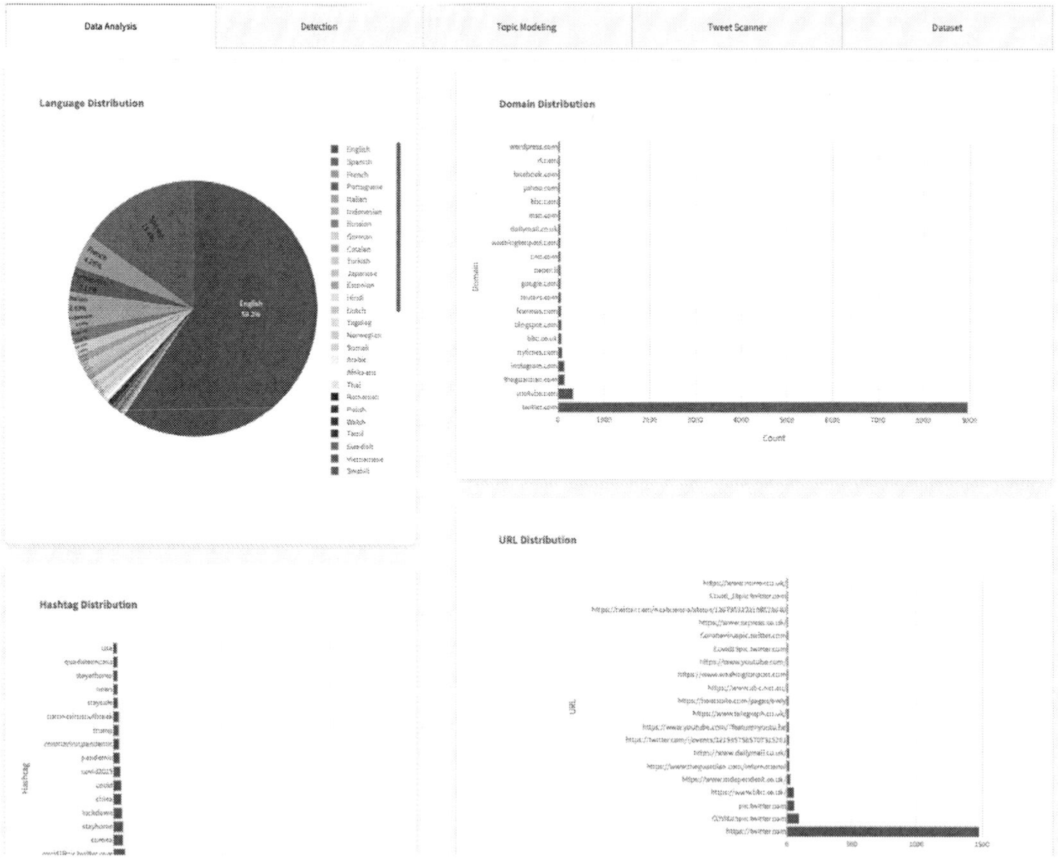

Figure 6: This screenshot of our web-platform shows the subpage for statistics of the dataset. The distribution of languages, domains, hashtags, and URLs are displayed.

Conversation-Aware Filtering of Online Patient Forum Messages

Anne Dirkson
LIACS, Leiden University
Niels Bohrweg 1, 2333 CA,
Leiden, the Netherlands

Suzan Verberne
LIACS, Leiden University
Niels Bohrweg 1, 2333 CA
Leiden, the Netherlands

Wessel Kraaij
LIACS, Leiden University
Niels Bohrweg 1, 2333 CA
Leiden, the Netherlands

`{a.r.dirkson, s.verberne, w.kraaij}@liacs.leidenuniv.nl`

Abstract

Previous approaches to NLP tasks on online patient forums have been limited to single posts as units, thereby neglecting the overarching conversational structure. In this paper we explore the benefit of exploiting conversational context for filtering posts relevant to a specific medical topic. We experiment with two approaches to add conversational context to a BERT model: a sequential CRF layer and manually engineered features. Although neither approach can outperform the F_1 score of the BERT baseline, we find that adding a sequential layer improves precision for all target classes whereas adding a non-sequential layer with manually engineered features leads to a higher recall for two out of three target classes. Thus, depending on the end goal, conversation-aware modelling may be beneficial for identifying relevant messages. We hope our findings encourage other researchers in this domain to move beyond studying messages in isolation towards more discourse-based data collection and classification. We release our code for the purpose of follow-up research.[1]

1 Introduction

In the past decade, social media has emerged as a source of valuable knowledge in the health domain (Gonzalez-Hernandez et al., 2017), for instance during the COVID-19 pandemic (Sarker et al., 2020) (Klein et al., 2020). In order to use social media to answer a medical question, it is necessary to identify posts on the forum that are relevant to the question at hand e.g. posts mentioning adverse drug responses (ADRs) (Li et al., 2020), personal experiences (Dirkson et al., 2019), medication abuse (Sarker et al., 2016) or medical misinformation (Kinsora et al., 2017). This filtering step is often the first step of the analysis pipeline. In this paper, we will refer to this specific type of filtering as relevance classification.

Previous automatic methods for medical relevance classification generally consider posts as units without context, thereby ignoring any information that can be gained from conversational context. One example of such an approach is the recent shared task on ADR relevance classification (Weissenbacher et al., 2019). Yet, including the conversational context may prove beneficial to relevance classification, as responses in a thread often relate to previous responses. For example, responses to a question or comment about a specific side effect are likely to also concern this side effect. To test this hypothesis, we investigate how positive labels are distributed across and within conversational threads.

At present, only one study into medical relevance classification has included some engineered features to capture aspects of the conversational structure (Kinsora et al., 2017). However, as this study includes only two discourse-based features, the effect of including manually engineered features that capture conversational structure is still largely unknown for relevance classification tasks.

Furthermore, including the relation between posts on a discourse level may also be able to improve classifier performance. Each post serves a conversational function in a dialogue, e.g. a question, explanation or statement (Austin, 1962). These functions are called *dialogue acts* (Stolcke et al., 2000). We have not found any study that included dialogue acts as features for medical relevance classification.

Proceedings of the 5th Social Media Mining for Health Applications (#SMM4H) Workshop & Shared Task, pages 11–18
Barcelona, Spain (Online), December 12, 2020.

As an alternative to using manually engineered features, conversational threads can also be modelled with a sequential model. This has proven beneficial in other fields such as rumor classification in social media discussions (Zubiaga et al., 2018). As of yet, the use of sequential models for medical relevance classification has also not been explored.

We address the following research questions in this paper:

RQ1 To what extent can the addition of a sequential model on top of state-of-the-art non-sequential models improve medical relevance classification of social media data?

RQ2 To what extent can the addition of manually engineered features for conversational structure and discourse improve medical relevance classification?

We use two different datasets for answering our questions. In our current research, we are particularly interested in discovering ADRs in online discussions. We have collected and annotated a dataset about this topic. Since this dataset is new, no other results have been published for it. We therefore use one other dataset for evaluating our methods: the medical misinformation dataset by Kinsora et al. (2017). We use a BERT-based model as baseline. BERT models constitute the current state of the art for most NLP tasks (Devlin et al., 2019) including ADR relevance classification (Weissenbacher et al., 2019).

In the following section, we will elaborate on related work. Hereafter, we describe our methodology and data in Section 3 and 4 respectively. Finally, we present and discuss our results in Section 5 and 6.

2   Related Work

The use of conversational structure for improving the performance of classifiers of social media posts is prevalent in the field of rumor classification (Zubiaga et al., 2018) and related fields like disagreement detection (Rosenthal and McKeown, 2015). Conversational structure has previously been exploited through (a) manually engineered features or (b) sequential classifiers.

The most commonly employed engineered features to model the conversational structure are the similarity to the previous message and to the thread in general (Zubiaga et al., 2018). In addition to these features, the current state-of-the-art model on a leading shared task for rumor stance classification (RumourEval-2019) uses the label of the previous message and the distance to the start of the thread (Li et al., 2019). In the health domain, the only study that employs manually engineered features for conversational structure is Kinsora et al. (2017). Specifically, they use the running count of positive labels and the distance to the previous positive label. In this study, we will employ the above features as well as expand upon them with additional discourse-related features.

Other studies have used sequential classifiers to model the discursive nature of social media, although according to Zubiaga et al. (2018) this is "still in its infancy" (p. 276). Their comparison of various classifiers for rumor stance classification revealed that sequential classifiers outperform non-sequential classifiers overall. This is probably due to their ability to leverage information about sequential structure and preceding labels. Furthermore, Zubiaga et al. (2018) found that sequential classifiers did not benefit from contextual features representing thread context (e.g. similarity to the source tweet) whereas non-sequential classifiers did. They speculate that sequential classifiers take the surrounding context into account implicitly. To see if this also holds true for relevance classification in medical social media, we will compare the addition of conversation-aware features to both sequential and non-sequential models.

3   Methods

3.1   Models

CRF   As a sequential model we use Conditional Random Fields (CRF). We train the models using the implementation in sklearn-crfsuite. L1 and L2 regularization parameters were tuned for each fold.

Linear SVM   As a non-sequential counterpart, we use the sklearn implementation of Linear Support Vector Machines. The hyper-parameter C is tuned per fold with a grid of 10^{-3} to 10^3 in steps of $\times 10$.

Feature type	Name	Description	Explanation (if applicable)
Local	+Emb	Sentence Vectors	We use Universal Sentence Encoder (USE) (Cer et al., 2018) to encode sentences into 512 dimensional vectors based on pre-trained embeddings so their cosine similarity (normalized between 0 and 1) approximates their semantic similarity.[2]
	+BERTpred	distilBERT predictions	The raw confidence scores for each label
Relational	+PrevSim	Similarity to previous message	Similarity is calculated using the USE sentence vectors
	+ThreadSim	Thread similarity	Similarity to USE sentence vector of all other posts in the thread combined into one vector
Positional	+Dist	Absolute distance from start of thread	
Label distribution	+PrevLbl	Label of previous post	We use the true labels for training and the predicted labels for testing for all label distribution features.
	+CountPos	Absolute running count of preceding positive labels in thread	
	+CountNeg	Absolute running count of preceding negative labels in thread	
	+RelPos	Percentage of preceding positive labels	
	+DistPos	Distance from previous positive label	
	+DistNeg	Distance from previous negative label	
Discourse	+DA	Dialogue act of post	Dialogue acts are calculated using the Dialogue Act tagger as trained by Tortoreto et al. (2019)
	+PrevDA	Dialogue act of previous post	

Table 1: Manually engineered features to model conversational structure

DistilBERT As BERT model, we opt for DistilBERT (distilbert-base-uncased), which is a lighter, more computationally efficient variant of BERT (Sanh et al., 2019). We use the Huggingface implementation (Wolf et al., 2019) with the wrapper ktrain (Maiya, 2020) to train our models. The initialization seed is set to 1. We use the default learning rate of 5×10^{-5} and tune the number of epochs (3 or 4) per fold.

Ensemble models To investigate the benefit of adding a sequential model on top of the DistilBERT model, we experiment with a blending-based ensemble method: we input the raw confidence scores from DistilBERT for each label as features in a CRF model (i.e. CRF + BERTpred). We create an equivalent non-sequential baseline by using the same approach with an SVM (i.e. SVM + BERTpred).

3.2 Feature analysis

To explore the benefit of manually engineered features that capture thread context, we use step-wise greedy forward feature selection using the features in Table 1. For each step-wise iteration, we select the best feature to add to the model until the F_1 score no longer improves. We use 10-fold cross-validation in which per fold features are selected on the development data (10%) and tested on a held-out test set (10%). For a fair comparison, we keep folds and hyper-parameters the same as for the respective base model. Since the label distribution features could leak information, we omit these gold annotated features for evaluation. Instead, we perform an initial run without these features and use the resulting predictions to calculate them for the final evaluation.

3.3 Model comparison

We used 10-fold cross validation in all experiments. Instead of splitting per message, we split on whole discussion threads to ensure possible dependencies between posts do not bias the outcome. Statistical comparisons of model performance are done using Wilcoxon signed rank tests across the 10 folds. To avoid the multiple testing problem, we only compare the three best models – namely those with the highest F_1 score, precision and recall – to the BERT baseline.

4 Data

Data collection At present, there is only one publicly available medical relevance classification data set that includes the conversational structure: the Medical Misinformation Data set (Kinsora et al., 2017).

---

[2] We opt for USE instead of BERT embeddings, as cosine similarity cannot be applied directly to BERT embeddings

Data set	Target	#Posts	#Discussions	Median length	% Positive
Medical Misinformation Dataset (Kinsora et al., 2017)	Misinformation	1,566	78	8.0	15.0 %
ADR Discussions (In-house)	Adverse Drug Response (ADR) & Coping Strategies	4,195	527	6	22.9 % & 12.3%

Table 2: Statistics on the data sets. The ADR Discussions data set has two target classes.

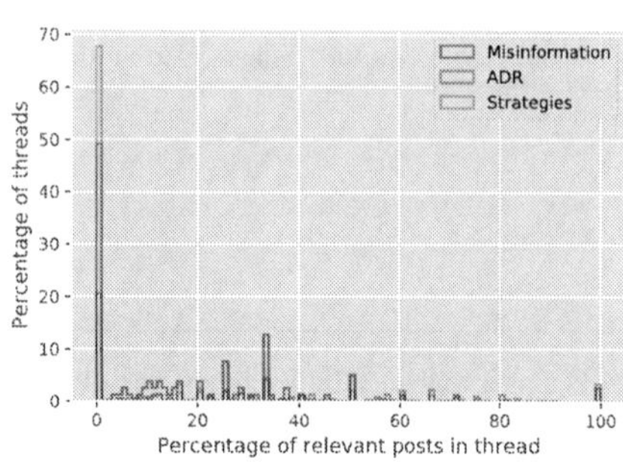

(a) Distribution of target posts *across* threads

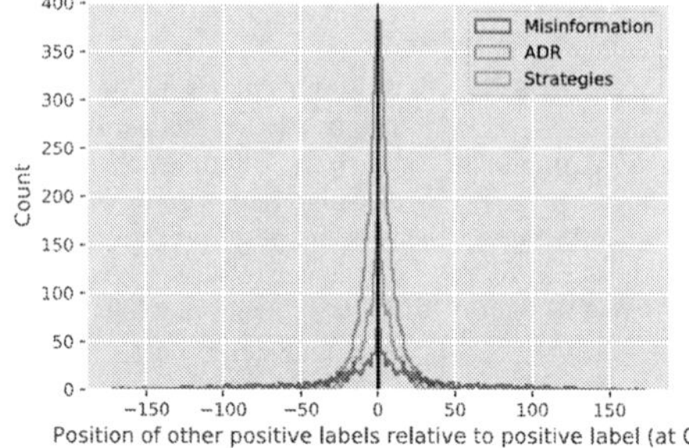

(b) The relative position of target posts to each other *within* threads

Figure 1: Distribution of the target class (i.e. positively labelled posts)

It is based on MedHelp data and annotated for the presence of misinformation. We collected a second data set from a Facebook group of Gastro Intestinal Stromal Tumor (GIST) patients. We selected 527 discussions based on their likelihood to contain an ADR: We selected the threads that contained (1) at least one drug name according to a match with RxNorm (U.S. National Library of Medicine, 2020) and (2) a high percentage of posts in which authors shared experiences. The latter criterion was included since sharing that you had an ADR is an example of experience sharing. To estimate this, we used a previously developed classifier (Dirkson et al., 2019). According to our classifier, at least 80% of the posts within each selected thread is a personal experience. Due to privacy issues and ownership of the data by the GIST International patient organization, we are not able to share this data set at present. See Table 2 for more details on the data sets.

Data annotation Following a pilot annotation round, the data was annotated by the first author and three patients for the presence of ADRs and coping strategies for dealing with ADRs (hereafter also called: Strategies) using an annotation guideline.[3] The pair-wise inter-annotator agreement was substantial for ADR (mean κ =0.71) and moderate for Coping Strategies (mean κ =0.54).

5   Results

5.1   Distribution of the target class in the discussion threads

As visualized in Figure 1a, the target class is not distributed equally across the discussion threads for any of the data sets; There appear to be many threads with few or no target posts. According to z-tests, the distribution is significantly different from normal. An inspection of the relative position of target posts *within* discussion threads reveals that target posts also cluster together (see Figure 1b). The probability that the post after a target post is also a target post is 27% for Misinformation and 40% and 34% for ADRs and Coping Strategies respectively. These probabilities are higher than is to be expected based on the percentage of positively labelled posts (see Table 2). Thus, it appears that the conversational structure is indeed related to the probability of a post being relevant and consequently incorporating conversational structure or discourse may be able to improve performance of relevance classifiers.

---

[3]Available at: `https://github.com/AnneDirkson/ConversationAwareFiltering`

	Misinformation		
	F_1	P	R
BERT	0.366 ± 0.155	0.386 ± 0.154	0.396 ± 0.235
SVM+Emb	**0.478 ± 0.083**	0.492 ± 0.109	0.482 ± 0.111
+ Features	0.392 ± 0.089	0.457 ± 0.169	0.405 ± 0.156
CRF+Emb	0.424 ± 0.155	**0.565 ± 0.148**	0.352 ± 0.162
+Features	0.457 ± 0.137	0.557 ± 0.155	0.420 ± 0.167
SVM + BERTpred	0.443 ± 0.078	0.449 ± 0.082	0.479 ± 0.151
+Features	0.454 ± 0.070	0.449 ± 0.081	**0.492 ± 0.140**
CRF + BERTpred	0.434 ± 0.079	0.453 ± 0.100	0.447 ± 0.138
+Features	0.428 ± 0.078	0.435 ± 0.092	0.446 ± 0.126

(a) Misinformation data set

	ADR				Strategies		
	F_1	P	R		F_1	P	R
BERT	**0.714 ± 0.034**	0.715 ± 0.038	0.718 ± 0.062	**BERT**	**0.581 ± 0.060**	0.622 ± 0.087	**0.563± 0.111**
SVM+Emb	0.640 ± 0.054	0.673 ± 0.055	0.613 ± 0.069	**SVM+Emb**	0.517 ± 0.101	**0.660 ± 0.111**	0.434 ± 0.111
+Features	0.610 ± 0.068	0.621 ± 0.087	0.624 ± 0.128	*+Features*	0.502 ± 0.108	0.603 ± 0.137	0.453 ± 0.128
CRF+Emb	0.654 ± 0.059	0.710 ± 0.036	0.611 ± 0.086	**CRF+Emb**	0.441 ± 0.134	0.597 ± 0.120	0.373 ± 0.151
+Features	0.638 ± 0.067	0.695 ± 0.037	0.601 ± 0.110	*+Features*	0.512 ± 0.106	0.609 ± 0.110	0.462 ± 0.143
SVM + BERTpred	**0.714 ± 0.035**	0.724 ± 0.043	0.707 ± 0.056	**SVM+Bertpred**	0.578 ± 0.059	0.632 ± 0.091	0.545 ± 0.089
+Features	0.677 ± 0.121	0.673 ± 0.164	**0.738 ± 0.103**	*+Features*	0.561 ± 0.095	0.601 ± 0.146	0.552 ± 0.087
CRF+ BERTpred	**0.714 ± 0.038**	**0.728* ± 0.040**	0.704 ± 0.062	**CRF + BERTpred**	**0.581 ± 0.065**	0.629 ± 0.087	0.558 ± 0.115
+Features	0.713 ± 0.039	0.726 ± 0.040	0.705 ± 0.060	*+Features*	0.573 ± 0.058	0.635 ± 0.090	0.539 ± 0.100

(b) ADR and Strategies data set

Table 3: Evaluation results of mean model performance over 10 folds. Features are selected through step-wise greedy feature selection. ****<0.01 *<0.05**

5.2 Model comparison

The results of model evaluation are presented in Table 3. It appears that neither the addition of a sequential layer nor manual features can improve upon the F_1 score of the BERT model. Misinformation detection appears to be the exception to this; any additional layer, sequential or not, outperforms the BERT baseline model. The highest F_1 is attained by an SVM model based on USE sentence vectors (+Emb), which were specifically designed for representing whole sentences. Perhaps sentence vectors perform better than BERT embeddings when the BERT model performs poorly (F_1= 0.366). Additional research will be necessary to substantiate this.

Despite a lack of improvement in the F_1 score for the detection of ADR and Strategies, an additional layer does seem to offer flexibility in tailoring the model towards a higher recall or precision. On the one hand, recall can be improved for two target classes by adding a non-sequential SVM layer with manual features to the BERT model. On the other hand, precision can be improved through the addition of a sequential CRF layer on top of BERT predictions for all target classes. Adding manually engineered features in addition to the sequential layer only improves the precision further for the detection of coping strategies. Our findings are thereby in line with Zubiaga et al. (2018). They speculated that sequential classifiers may take the surrounding context into account implicitly and therefore do not benefit from features representing thread context.

The only significant increase according to Wilcoxon signed rank tests is in the precision for ADR detection. This may be related to the high variance between folds. Further research is necessary to validate these results and advance our understanding of how conversation-aware modelling can be best be used for relevance classification. We believe that this first study shows that this is a promising direction.

5.3 Analysis of selected features

There is large variation in which features are selected per fold. Manual inspection of the selected features shows that features relating to the distribution of labels in the thread are chosen most often, especially the running count of negative and positive labels in the thread (CountNeg, CountPos), and the label of

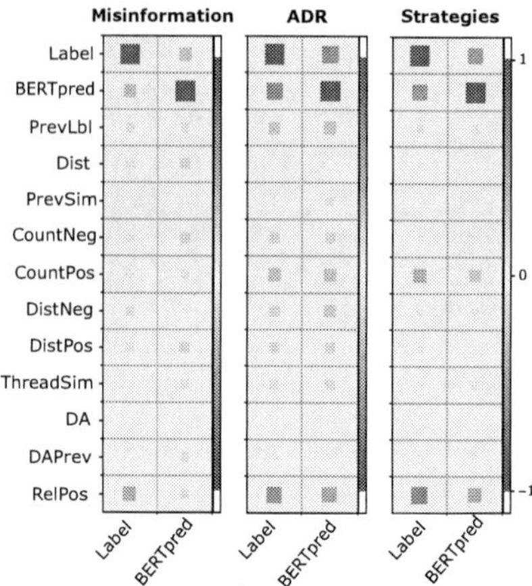

Figure 2: Correlation matrix of ground truth labels and BERT predictions with the manually engineered features. The size and colour of the squares corresponds to the strength of the correlation

the previous post (PrevLbl) (see Table 1). Features of this type may therefore be the most promising for future work. The number of features that is chosen is more consistent; On average, 1 or 2 of the 11 features are chosen.

To further explore why certain features are chosen, we compute the correlations between the target label and the manually engineered features and between the BERT predictions and the manually engineered features (see Figure 2). We find, firstly, that features relating to the label distribution indeed appear to correlate most strongly with the ground truth labels. Secondly, the correlation between these features and the BERT predictions is often equal to or stronger than the respective correlation to the ground truth. This might indicate that this variance is already captured by the BERT model and therefore manually engineered features have little to add to the baseline model.

6 Discussion

We find that the distribution of target posts across discussion threads is skewed and that within a conversational thread posts cluster together. Thus, our hypothesis that the probability of a target post occurring is related to the conversational structure appears valid.

In answer to **RQ1**, we find that a sequential CRF layer on top of a BERT model improves precision slightly, although only significantly so for ADR detection. In answer to **RQ2**, we find that the addition of manually engineered features representing thread context often does not aid performance. The one consistent exception is when combined with a non-sequential SVM layer on top of a BERT model. This combination can improve recall for all target classes, although not significantly. An additional layer on top of a BERT model that is able to capture the thread context appears to offer flexibility in tailoring the model towards a higher recall or precision. In future work, we plan to investigate the benefit of including conversational context for other tasks such as concept normalization of ADR.

For all the data sets included in this study, a pre-selection of discussion threads was made prior to annotation to ensure a higher proportion of target posts. We expect that both sequential models and manually engineered features of thread context may prove more beneficial when such a pre-selection does not take place and the target class is even more imbalanced. Thus, our results may be an underestimation of the benefit of conversational context for finding 'needles in the haystack'.

Finally, our findings call into question the practice of splitting data into folds without taking the discussion context into account. In this study, we split the folds per discussion thread and we recommend others to consider doing so when dealing with multiple posts from the same thread, as neglecting to do so when there are dependencies between posts may bias model performance. This is especially important when threads contain duplicate posts.

Acknowledgements

This work was funded by the SIDN fonds. We would like to thank our annotators for their time and effort and our anonymous reviewers for their feedback.

References

John Langshaw Austin. 1962. *How to do things with words.* Oxford university press.

Daniel Cer, Yinfei Yang, Sheng-yi Kong, Nan Hua, Nicole Limtiaco, Rhomni St. John, Noah Constant, Mario Guajardo-Cespedes, Steve Yuan, Chris Tar, Brian Strope, and Ray Kurzweil. 2018. Universal sentence encoder for English. In *Proceedings of the 2018 Conference on Empirical Methods in Natural Language Processing: System Demonstrations*, pages 169–174, Brussels, Belgium, November. Association for Computational Linguistics.

Jacob Devlin, Ming-Wei Chang, Kenton Lee, and Kristina Toutanova. 2019. Bert: Pre-training of deep bidirectional transformers for language understanding. In *NAACL-HLT*.

Anne Dirkson, Suzan Verberne, and Wessel Kraaij. 2019. Narrative Detection in Online Patient Communities. In A. Jorge, R. Campos, A. Jatowt, and S. Bhatia, editors, *Proceedings of the Text2StoryIR'19 Workshop*. CEUR-WS.

Graciela Gonzalez-Hernandez, Abeed Sarker, Karen O 'Connor, and Guergana Savova. 2017. Capturing the Patient's Perspective : a Review of Advances in Natural Language Processing of Health-Related Text. *Yearbook of medical informatics*, pages 214–217.

Alexander Kinsora, Kate Barron, Qiaozhu Mei, and Vinod Vydiswaran. 2017. Creating a Labeled Dataset for Medical Misinformation in Health Forums. In *IEEE International Conference on Healthcare Informatics*.

Ari Klein, Arjun Magge, Karen O'Connor, Haitao Cai, Davy Weissenbacher, and Graciela Gonzalez-Hernandez. 2020. A chronological and geographical analysis of personal reports of covid-19 on twitter. *medRxiv (preprint)*.

Quanzhi Li, Qiong Zhang, and Luo Si. 2019. eventAI at SemEval-2019 Task 7: Rumor Detection on Social Media by Exploiting Content, User Credibility and Propagation Information. In *Proceedings of the 13th International Workshop on Semantic Evaluation (SemEval-2019)*, pages 855–859.

Zhiheng Li, Zhihao Yang, Ling Luo, Yang Xiang, and Hongfei Lin. 2020. Exploiting adversarial transfer learning for adverse drug reaction detection from texts. *Journal of Biomedical Informatics*, page 103431.

Arun S. Maiya. 2020. ktrain: A low-code library for augmented machine learning. *arXiv*, arXiv:2004.10703 [cs.LG].

Sara Rosenthal and Kathy McKeown. 2015. I couldn't agree more: The role of conversational structure in agreement and disagreement detection in online discussions. In *Proceedings of the 16th Annual Meeting of the Special Interest Group on Discourse and Dialogue*, pages 168–177, Prague, Czech Republic, September. Association for Computational Linguistics.

Victor Sanh, Lysandre Debut, Julien Chaumond, and Thomas Wolf. 2019. DistilBERT, a distilled version of BERT: smaller, faster, cheaper and lighter. In *5th Workshop on Energy Efficient Machine Learning and Cognitive Computing - NeurIPS 2019*.

Abeed Sarker, Karen O'connor, Rachel Ginn, Matthew Scotch, Karen Smith, Dan Malone, and Graciela Gonzalez. 2016. Social Media Mining for Toxicovigilance: Automatic Monitoring of Prescription Medication Abuse from Twitter. *Drug Safety*, 39.

Abeed Sarker, Sahithi Lakamana, Whitney Hogg-Bremer, Angel Xie, Mohammed Ali Al-Garadi, and Yuan-Chi Yang. 2020. Self-reported COVID-19 symptoms on Twitter: an analysis and a research resource. *Journal of the American Medical Informatics Association*, 27(8):1310–1315, 07.

Andreas Stolcke, Klaus Ries, Noah Coccaro, Elizabeth Shriberg, Rebecca Bates, Daniel Jurafsky, Paul Taylor, Rachel Martin, Carol Van Ess-Dykema, and Marie Meteer. 2000. Dialogue act modeling for automatic tagging and recognition of conversational speech. *Computational Linguistics*, 26(3):339–374.

Giuliano Tortoreto, Evgeny A Stepanov, Alessandra Cervone, Mateusz Dubiel, and Giuseppe Riccardi. 2019. Affective Behaviour Analysis of On-line User Interactions: Are On-line Support Groups more Therapeutic than Twitter? In *Proceedings of the 4th Social Media Mining for Health Applications (#SMM4H) Workshop & Shared Task*, pages 79–88.

U.S. National Library of Medicine. 2020. RxNorm.

Davy Weissenbacher, Abeed Sarker, Arjun Magge, Ashlynn Daughton, Karen O'Connor, Michael Paul, and Graciela Gonzalez-Hernandez. 2019. Overview of the Fourth Social Media Mining for Health (#SMM4H) Shared Task at ACL 2019. In *Proceedings of the 4th Social Media Mining for Health Applications (#SMM4H) Workshop & Shared Task*, pages 21–30.

Thomas Wolf, Lysandre Debut, Victor Sanh, Julien Chaumond, Clement Delangue, Anthony Moi, Pierric Cistac, Tim Rault, Rémi Louf, Morgan Funtowicz, and Jamie Brew. 2019. Transformers: State-of-the-art Natural Language Processing. *ArXiv*.

Arkaitz Zubiaga, Elena Kochkina, Maria Liakata, Rob Procter, Michal Lukasik, Kalina Bontcheva, Trevor Cohn, and Isabelle Augenstein. 2018. Discourse-Aware Rumour Stance Classification in Social Media Using Sequential Classifiers. *Information Processing & Management*, 54(2):273–390.

Annotating Patient Information Needs in Online Diabetes Forums

Julia Romberg[1], Jan Dyczmons[2,3,4], Sandra Olivia Borgmann[2,3,4]
Jana Sommer[2,3,4], Markus Vomhof[2,3,4], Cecilia Brunoni[2,3,4]
Ismael Bruck-Ramisch[2,3,4], Luis Enders[2,3,4]
Andrea Icks[2,3,4], Stefan Conrad[1]

[1]Institute of Computer Science, Heinrich Heine University Düsseldorf
[2] Institute for Health Services Research and Health Economics, German Diabetes Center (DDZ),
Leibniz Center for Diabetes Research at the Heinrich Heine University Düsseldorf
[3]Institute for Health Services Research and Health Economics, Centre for Health and Society,
Faculty of Medicine, Heinrich Heine University Düsseldorf
[4]German Center for Diabetes Research (DZD), München-Neuherberg, Germany
`julia.romberg@hhu.de`

Abstract

Identifying patient information needs is an important issue for health care services and implementation of patient-centered care. A relevant number of people with diabetes mellitus experience a need for information during the course of the disease. Health-related online forums are a promising option for researching relevant information needs closely related to everyday life. In this paper, we present a novel data corpus comprising 4,664 contributions from an online diabetes forum in German language. Two annotation tasks were implemented. First, the contributions were categorised according to whether they contain a diabetes-specific information need or not, which might either be a non diabetes-specific information need or no information need at all, resulting in an agreement of 0.89 (Krippendorff's α). Moreover, the textual content of diabetes-specific information needs was segmented and labeled using a well-founded definition of health-related information needs, which achieved a promising agreement of 0.82 (Krippendorff's α_u). We further report a baseline for two sub-tasks of the information extraction system planned for the long term: contribution categorization and segment classification.

1 Motivation

Diabetes mellitus (DM) is a chronic disease with an estimated global prevalence of about 10% and steadily rising (Saeedi et al., 2019). To reduce the risk of diabetes-related acute and long-term complications, continuous health care and support as well as a high level of self-management skills are required (American Diabetes Association, 2020). A relevant number of people with DM experience a need for information during the course of the disease (Grobosch et al., 2018; Biernatzki et al., 2018). According to Ormandy (2011), a patient information need is defined as "the recognition that their knowledge is inadequate to satisfy a goal, within the context or situation that they find themselves at a specific point in the time". For example, shortly after diagnosis, individuals with DM report a high need for information with regard to new treatment strategies that simplify their everyday life (Grobosch et al., 2018).

To support individuals with DM adequately, identifying their information needs is an important issue in the implementation of patient-centered care (Scholl et al., 2014). Nevertheless, research in this area is still sparse. Only a few studies consider information needs during the course of the disease or in everyday life (Grobosch et al., 2018; Biernatzki et al., 2018). Studies from the field of diabetes often rely on classic quantitative or qualitative survey methods, such as questionnaires and interviews. However, to find the information they need, people with DM use online health communities, for example online diabetes forums, to improve their disease management (VanDam et al., 2017; Kuske et al., 2017; Reidy et al., 2019). An online forum can be described as an "asynchronous communication with other community members (e.g., patients can put questions to professionals or peers)" (van der Eijk et al., 2013).

We focus on the analysis of textual contributions from online diabetes forums, as personal experiences reported with classic methods may differ from personal experiences reported in online forums. An

Proceedings of the 5th Social Media Mining for Health Applications (#SMM4H) Workshop & Shared Task, pages 19–26
Barcelona, Spain (Online), December 12, 2020.

advantage compared to traditional methods is that the communication of people with DM in online diabetes forums can highlight information needs that are more closely related to everyday life (Seale et al., 2010). This might be due to the perceived anonymity in online forums and the absence of a researcher, which can result in an openness of the forum users (Jamison et al., 2018). Nevertheless, there are several challenges for researchers using online forums as data sources. These include identifying relevant passages in a mass of posts (VanDam et al., 2017) and consideration of the more spoken language used by users when writing their posts (Seale et al., 2010). To the best of our knowledge, there is only one previous study in the field of information needs in diabetes, which uses online forums as data sources (Biernatzki et al., 2018): Ravert et al. (2004) analysed posts in online forums from adolescents with type 1 DM. They found that adolescents with type 1 DM visit online forums primarily to seek for social support, to receive information or counseling, and to exchange experiences. These findings support our motivation to explore information needs in the context of online forums.

Machine learning and natural language processing have been increasingly applied for the research of health-related issues in social media, e.g. content analysis, opinion mining, and the development of domain-specific lexicons (Denecke and Nejdl, 2009; Chen, 2012; Sudau et al., 2014; Sokolova and Bobicev, 2013; Bobicev et al., 2012; Goeuriot et al., 2012). There are two recent works focusing in particular on diabetes (Liu and Chen, 2015; Bell et al., 2018). Liu and Chen (2015) built a system for the detection and extraction of adverse drug events for drug safety surveillance. The developed approaches are evaluated on a case study corpus from a diabetes patient forum in the United States. Bell et al. (2018) investigated the automatic detection of the risk of type 2 DM directly from the Twitter activity of a person. In both cases English language data sets are used.

In this paper we present a novel corpus of German-language forum data on diabetes in which information needs are coded. The corpus was created with two objectives in mind, namely (a) the recognition of diabetes-specific information needs at contribution level and (b) the identification of relevant text segments within contributions that contain a diabetes-specific information need. Our intermediate research goal is the building of an information extraction (IE) system which consists of three steps. First, contributions with diabetes-specific information needs are filtered. Secondly, relevant text segments within the contributions with diabetes-specific information needs are identified. Third, the relevant text segments are classified according to the dimensions of patient information needs. The developed data corpus will serve as training set for the development of such a system. We also show a first approach to both classification tasks. The text segmentation step will be investigated in future work. In the long term, the IE system will allow for an automatic analysis with regard to information needs in online diabetes forums, e.g. which topics trigger increased information needs and which groups have a need, and to develop a search system that can help individuals finding relevant forum posts related to their diabetes-specific information needs.

The remainder of the paper is structured as follows. Section 2 introduces the definition of patient information needs on which we base our annotation schema. Subsequently, Section 3 outlines the data basis for the corpus, the annotation process is described in detail and the resulting corpus is analysed. Finally, Section 4 presents a basic approach for the contribution categorization and the segment classification tasks and Section 5 concludes our work and highlights the future research.

2 Ormandy's Definition of Patient Information Needs

Ormandy (2011) defines a patient information need as the "recognition that their knowledge is inadequate to satisfy a goal, within the context/situation that they find themselves at a specific point in the time". Within this definition, the following four concepts are addressed, which can either activate or influence information needs: (i) goal/purpose, (ii) context, (iii) situation and (iv) time.

(i) **Goal/Purpose:** A need for information arises from the underlying purpose of trying to achieve one or more goals, e.g. to perform self-management tasks.

(ii) **Context:** Context factors are defined according to the Wilson model of information behaviour as factors that influence the process of information seeking (Wilson, 2000). These include psychological and cognitive factors (e.g. emotions and interests), stress and coping strategies, perceived self-efficacy, demographic factors (e.g. age and gender), and role-related and environmental factors (e.g. social networks and access to health care).

(iii) **Situation:** The situation is defined as the "particular set of circumstances in which people find themselves that creates an awareness of an information need" (Ormandy, 2011). Possible situations can be events, experiences, or encounters. Examples of health-related situations are the perception of symptoms or even the perception of a life-threatening condition.

(iv) **Time:** Information needs arise at a certain point in time during an individual's progress of disease which can be influenced by context and situation factors.

3 Data Corpus

In the following, the data source and the annotation process are presented. In addition, the reliability of the annotations is measured and the resulting data set is analysed.

3.1 Source and Preprocessing

The data corpus was constructed from the German-language online forum *forum.diabetesinfo.de*, which appears to be of reasonably high linguistic quality compared to other forums focusing on diabetes. Only publicly accessible threads were considered. When registering in the forum, forum users were made aware by the forum operator that these posts are made visible to the public. Contributions are written using pseudonyms, thus preventing direct inference to a person. Under the assumption that information needs are primarily formulated in the initial post of a thread, only the first contributions and the thread titles were included in the data set. Foreign-language posts and contributions shorter than 20 characters were excluded. HTML tags, emoticons, pictures, and applications were filtered out. Links, quotations, and tables were retained. After preprocessing, a share of 4,664 documents was used to build the annotated corpus. Additional 557 documents were used to develop the annotation guidelines.

3.2 Annotation Tasks

In order to create a suitable training data set for the information extraction system, two aspects must be taken into account. First, the documents of the corpus need to be coded with regard to the existence of a diabetes-specific need for information. Second, the textual content of documents in which a diabetes-specific information need is described must be segmented and the units of information must be labeled according to Ormandy's definition.

Annotation Task 1: Document Categorization

The data set consists of documents with diverse content. Not every contribution was written to satisfy an information need, which makes it necessary to categorize the documents according to whether information needs are included or not. Contributions that fall into the latter category are, for instance, experience reports or success stories that are shared with the community. Review of the data also revealed that some information needs are not related to diabetes. Contributions with diabetes-specific information needs range from technical questions, e.g. on pumps, to nutritional issues. Typical postings for non-diabetes-specific information needs are e.g. questions about forum functions. Since our research focus is on diabetes-specific information needs, a further subdivision is made. This results in an annotation scheme on document-level with three categories: *diabetes-specific information need* (abbreviated IN_{ds}), *non-diabetes-specific information need* (abbreviated IN_{oth}) and *no information need* (abbreviated None). Figure 1 shows exemplary English-language contributions for each of the three categories.

Annotation Task 2: Detection and Assignment of the Dimensions of an Information Need

If a document contains a diabetes-specific information requirement, the IE system to be developed should extract the relevant information. In order to train this component, segments expressing the different dimensions of an information need must first be coded in the data corpus. To this end, we first discuss under which restrictions the definition of Ormandy can be transferred to health-related online forums. Analysis of the present corpus shows that the concepts goal/purpose, context and situation are represented

Hey everyone, just wanted to tell you that I have been trying the diet. I feel so much better now. Really proud of me!	Hi, I got a problem. Before breakfast my blood sugar is ok. Next, I inject insulin and have breakfast. An hour later my blood sugar is really low. I am seriously worried, why is this happening and what can I do about it?	This is not a question about diabetes, but maybe you can help me anyway. When I tried to transfer my data from my mobile to my PC, I deleted everything. How do I get my data back?
(a) None	(b) IN_{ds}	(c) IN_{oth}

Figure 1: Exemplary contributions with document categories (a,b,c) and labeled (context/situation, goal/situation) segments of a diabetes-specific information need (b).

in the contributions. For the temporal aspect of an information need, however, this is not the case. In most cases the writer does not refer to the time her reported information need evolved. A further problem arises in the clear separation of context and situation. During the development of the guidelines it became apparent that the annotators often chose similar units for the two categories, but they often did not agree whether the dimension should be labeled as context or situation. Ormandy (2011) mentions: "A term closely related to context is situation, usually used with a narrower meaning (...)". Since the categories cannot be clearly separated, they are hence grouped into one dimension. It is also necessary to consider how the text spans to be coded can be determined. Initially, we considered using sentences as fixed units. After reflecting on the data source, it became apparent that often different concepts are expressed in separate parts of the same sentence. As a consequence, we decided to allow the annotators to freely divide text segments on token-level within sentence boundaries. Concluding, this leads to a text segmentation task in which the resulting text segments are assigned to one of two dimensions: *goal/purpose* and *context/situation*. Figure 1b shows an annotation example in English.

3.3 Annotation Setup

Annotation guidelines were developed in sub-steps on the basis of 557 documents from the same online diabetes forum. The annotation was performed with the brat rapid annotation tool (Stenetorp et al., 2012). Following the developed guidelines, a share of 750 documents of the total 4,664 corpus documents was annotated by three coders to calculate the agreement and then adjudicated by a supervising person to build a gold standard. Due to limited capacity we decided to have the remaining 3,914 documents, hereinafter called silver standard, processed by respectively one of the three trained annotators without adjudication. However, the jointly coded part allows us to get a good estimate of the agreement between the coders. Annotations are available for download at https://dbs.cs.hhu.de/datasets/diabetesanno/.

3.4 Inter-Annotator Agreement

The reliability of annotation task 1 was measured using Fleiss's κ (Fleiss, 1971) and Krippendorff's α (Krippendorff,). The κ measure determines the agreement assuming that all categories are equally dissimilar. To also take into account the intuitive assumption that contributions in both categories on existing information needs are more similar to each other than to contributions without any information need, we used weighted α using the following distance function for two codings c and k:

$$d(c,k) = \begin{cases} 0, & c = k \\ 1, & c \neq k \text{ and } c,k \in \{\text{IN}_{ds}, \text{IN}_{oth}\} \\ 2, & else \end{cases}$$

Coding the documents with the three-class scheme resulted in high agreement with a Fleiss's κ coefficient of 0.86 and a Krippendorff's α coefficient of 0.89.

Annotation task 2 likewise scored a solid agreement. We have measured the reliability of the segmentation with Krippendorff's α_u (Krippendorff, 1995), a coefficient for unitizing tasks. In total, an average agreement of 0.82 per document was measured. The two dimensions individually showed an average α_u of 0.82 (goal/purpose) and an average α_u of 0.79 (context/situation).

3.5 Corpus Analysis

For annotation task 1, 4,664 contributions have been coded. A single document consists of 127 tokens on average. Table 1 gives an overview of the resulting data set. In gold and silver standard, with

category	gold standard	silver standard	total
IN_{ds}	323	1,844	2,167
IN_{oth}	35	173	208
None	392	1,897	2,289
total	750	3,914	4,664

Table 1: Distribution of the contributions among the categories for annotation task 1.

dimension	gold standard	silver standard	total
context/situation	1,867	10,877	12,744
goal/purpose	672	3,679	4,351
total	2,539	14,556	17,095

Table 2: Distribution of the text segments among the dimensions for annotation task 2.

378 of 750 contributions and 2,017 of 3,914 contributions respectively, information needs are equally expressed in about half of all documents. Within these documents, the proportion of contributions containing information needs related to diabetes amounts to around 90 percent in each case. The 2,167 documents with diabetes-specific information needs taken into account in annotation task 2 consist of approximately 300,000 token, of which about 80 percent were assigned to some text segment containing either goal/purpose or context/situation. Goal/purpose mentions are fundamental for a need for information and could be identified in each document, whereas context or situation information is present in 2,044 contributions. A summary of the data set is quantified in Table 2. 17,095 text segments have been identified overall, of which about 35 percent express a goal or purpose and about 65 percent consist of a contextual or situational statement. This distribution applies to both the gold and the silver part of the corpus.

Some particularities can be found in the data corpus. The user-generated contributions show a high number of spelling mistakes, missing or unintuitive choice of punctuation and lack of semantic coherence within the texts. In addition, forum users use a very unique vocabulary, including forum-specific abbreviations for diabetes-related terms. Technical issues also seem to be central to the forum users, as technical aids (devices or software) play an important role for diabetes patients to manage and simplify their everyday life. As a result, the annotators sometimes had difficulty in understanding content, which made the annotations more challenging. In particular contributions to technical aids and medical specialities required additional knowledge to assess the extent to which these contributions are relevant to diabetes. This was also reflected in discrepancies by the annotators in the decision whether or not an information need is diabetes-specific. Some people write in a self-reflecting nature, implicitly raising information needs. Yet without an explicit statement of a goal/purpose, it is difficult to grasp the writer's intention, e.g. whether she seeks for support or not. A further difficulty in the annotation process was to decide which context and situation information is relevant to an information need.

The characteristics described above, in particular the semantic and syntactic challenges of the contributions as well as the required context knowledge and expertise for understanding them, also represent a special challenge for the IE system to be developed.

4 Baseline Approaches

In this section, we provide baselines (a) for the first step in the IE system, the contribution categorization problem, and (b) for the third step in the IE system, the classification of given text segments according to the dimensions of an information need. The intermediate segmentation step will be examined in future work. The baselines are intended to serve as a basis reference value for follow-up research. The method was the same for both tasks. Texts (either entire contributions or text segments) were split into lower case lemma tokens using *spaCy*[1] and *IWNLP* (Liebeck and Conrad, 2015) and stop words were removed. The classification was implemented with *scikit-learn* (Pedregosa et al., 2011). As classifier we chose a SVM, with fixed C of 1 and linear kernel, as SVMs are known to achieve good results for various text classification tasks. The evaluation was performed through a five-fold cross-validation. All unigrams that occurred in at least 10 documents and in at most 80 percent of the respective training data set were retained in the vocabulary, and the resulting representations were weighted with *tf-idf*.

Table 3 shows the results for both tasks. We first treat the contribution categorization task as a three-class problem. Diabetes-specific information needs are already recognised with a promising F_1 of 0.75.

---

[1]`https://spacy.io/`

	Contribution Categorization			Segment Classification	
	None	IN_{ds}	IN_{oth}	goal/purpose	context/situation
Precision	0.78	0.73	0.20	0.75	0.80
Recall	0.71	0.82	0.00	0.29	0.41
F_1	0.71	0.75	0.01	0.41	0.87
F_1 macro		0.49			0.64
F_1 micro		0.70			0.76

Table 3: Results for contribution categorization and segment classification.

The class IN_{oth} is poorly hit, which is likely due to the class imbalance in the data set. While None and IN_{ds} are about the same size, the class size of IN_{oth} with a total of 208 instances makes up less than 5% of all contributions. Analysis of wrong predictions between IN_{ds} and None shows that a frequent mistake occurs in the recognition of rather technical, often long reports describing and evaluating products (e.g. blood sugar measuring devices, pumps). These are often erroneously assigned to IN_{ds} although they do not contain information needs. This might be due to the subject matter, showing similar wording to some contributions with diabetes-specific information needs. The F_1 values of 0.75 for IN_{ds} and 0.71 for None nevertheless show that, despite a high degree of overlap in vocabulary, there are terms or term combinations that are class specific. Another interesting aspect relates to questions. Contrary to the expectation that explicit questions with a corresponding punctuation mark are a clear indication of the existence of an information need, this did not hold true. On the one hand, contributions with rhetorical or self-reflexive questions (primarily in narrative contributions), e.g. "How did I come across this again?", were recognised as IN_{ds}. On the other hand, quite unexpectedly, short contributions consisting mainly of goal/purpose questions were assigned None. This illustrates that the use of questions is not a specific evidence of a need for information in this data corpus. The error analysis showed that a SVM based on unigrams is insufficient for the task. Instead, a more sophisticated approach that involves the context of a contribution to decide whether words or sentences indicate information needs must be used. This difficulty becomes particularly evident in the case of questions which can take on different functions depending on the context in which they are found, e.g. expressing the goal/purpose of an information need or, conversely, giving no indication of an information need (rhetorical or self-reflexive questions).

When classifying text segments according to the dimensions of an information need, good precision values of 0.75 and 0.80 were achieved, but the recall rate was noticeably weaker. Nevertheless, the F_1 macro value of 0.64 shows that our model was able to learn at least some characteristics of both classes and goes beyond pure guessing. The results imply that the correct assignment of the two dimensions of an information need based on the wording of a single text segment is very difficult. The concepts of the single segments might only become clear from the overall context of all relevant segments of a contribution.

5   Conclusion and Future Work

With the long-term goal of developing an IE system for diabetes-specific information needs from forums, we introduced a German-language online diabetes forum corpus annotated on different levels. 4,664 forum posts were coded for the identification of diabetes-specific information needs on document-level. 2,167 contributions that contain a diabetes-specific information need were segmented and labeled using a well-founded definition. In this context we discussed how Ormandy's definition of patient information needs can be applied to the domain of online forums. The resulting agreement values of 0.86 κ, 0.89 α and 0.82 α_u prove good reliability. Further, a SVM approach was applied. A promising F_1 of about 0.75 was achieved in the identification of diabetes-specific information needs at contribution level. For the finer classification of text segments according to the dimensions of an information need, the approach performed weaker, with an improvable F_1 macro of 0.64. The results offer a baseline for further work.

Following this work, we will improve the classification approaches and also work on text segmentation methods, in order to develop a complete information extraction system in the long run.

References

American Diabetes Association. 2020. Introduction: Standards of Medical Care in Diabetes-2020. *Diabetes Care*, 43(Suppl 1):1–2.

Dane Bell, Egoitz Laparra, Aditya Kousik, Terron Ishihara, Mihai Surdeanu, and Stephen Kobourov. 2018. Detecting Diabetes Risk from Social Media Activity. In *Proceedings of the Ninth International Workshop on Health Text Mining and Information Analysis*, pages 1–11. Association for Computational Linguistics.

Lisa Biernatzki, Silke Kuske, Jutta Genz, Michaela Ritschel, Astrid Stephan, Christina Bächle, Sigrid Droste, Sandra Grobosch, Nicole Ernstmann, Nadja Chernyak, and Andrea Icks. 2018. Information needs in people with diabetes mellitus: a systematic review. *Systematic Reviews*, 7:27.

Victoria Bobicev, Marina Sokolova, Yasser Jafer, and David Schramm. 2012. Learning Sentiments from Tweets with Personal Health Information. In *Proceedings of the 25th CanadianConference on Advances in Artificial Intelligence*, pages 37–48. Springer, Berlin, Heidelberg.

Annie T. Chen. 2012. Exploring online support spaces: Using cluster analysis to examine breast cancer, diabetes and fibromyalgia support groups. *Patient Education and Counseling*, 87(2):250–257.

Kerstin Denecke and Wolfgang Nejdl. 2009. How valuable is medical social media data? Content analysis of the medical web. *Information Sciences*, 179(12):1870–1880.

Joseph L. Fleiss. 1971. Measuring nominal scale agreement among many raters. *Psychological Bulletin*, 76(5):378–382.

Lorraine Goeuriot, Jin-Cheon Na, Wai Yan Min Kyaing, Christopher Khoo, Yun-Ke Chang, Yin-Leng Theng, and Jung-Jae Kim. 2012. Sentiment Lexicons for Health-Related Opinion Mining. In *Proceedings of the 2nd ACM SIGHIT International Health Informatics Symposium*, pages 219–226. Association for Computing Machinery.

Sandra Grobosch, Silke Kuske, Ute Linnenkamp, Nicole Ernstmann, Astrid Stephan, Jutta Genz, Alexander Begun, Burkhard Haastert, Julia Szendroedi, Karsten Müssig, Volker Burkart, Michael Roden, and Andrea Icks. 2018. What information needs do people with recently diagnosed diabetes mellitus have and what are the associated factors? A cross-sectional study in Germany. *BMJ Open*, 8(10):e017895.

James Jamison, Stephen Sutton, Jonathan Mant, and Anna De Simoni. 2018. Online stroke forum as source of data for qualitative research: insights from a comparison with patients' interviews. *BMJ Open*, 8(3):e020133.

Klaus Krippendorff. Computing Krippendorff's Alpha-Reliability. *Annenberg School for Communication: Departmental Papers*. Retrieved from https://repository.upenn.edu/asc_papers/43.

Klaus Krippendorff. 1995. On the Reliability of Unitizing Continuous Data. *Sociological Methodology*, 25:47–76.

Silke Kuske, Tim Schiereck, Sandra Grobosch, Andrea Paduch, Sigrid Droste, Sarah Halbach, and Andrea Icks. 2017. Diabetes-related information-seeking behaviour: a systematic review. *Systematic Reviews*, 6:212.

Matthias Liebeck and Stefan Conrad. 2015. IWNLP: Inverse Wiktionary for Natural Language Processing. In *Proceedings of the 53rd Annual Meeting of the Association for Computational Linguistics and the 7th International Joint Conference on Natural Language Processing (Volume 2: Short Papers)*, pages 414–418. Association for Computational Linguistics.

Xiao Liu and Hsinchun Chen. 2015. Identifying Adverse Drug Events from Patient Social Media: A Case Study for Diabetes. *IEEE Intelligent Systems*, 30(3):44–51.

Paula Ormandy. 2011. Defining information need in health–assimilating complex theories derived from information science. *Health Expectations*, 14:92–104.

Fabian Pedregosa, Gaël Varoquaux, Alexandre Gramfort, Vincent Michel, Bertrand Thirion, Olivier Grisel, Mathieu Blondel, Peter Prettenhofer, Ron Weiss, Vincent Dubourg, Jake Vanderplas, Alexandre Passos, David Cournapeau, Matthieu Brucher, Matthieu Perrot, and Édouard Duchesnay. 2011. Scikit-learn: Machine learning in Python. *Journal of Machine Learning Research*, 12:2825–2830.

Russell D. Ravert, Mary D. Hancock, and Gary M. Ingersoll. 2004. Online Forum Messages Posted by Adolescents with Type 1 Diabetes. *The Diabetes Educator*, 30(5):827–834.

Claire Reidy, David C. Klonoff, and Katharine D. Barnard-Kelly. 2019. Supporting Good Intentions With Good Evidence: How to Increase the Benefits of Diabetes Social Media. *Journal of Diabetes Science and Technology*, 13(5):974–978.

Pouya Saeedi, Inga Petersohn, Paraskevi Salpea, Belma Malanda, Suvi Karuranga, Nigel Unwin, Stephen Colagiuri, Leonor Guariguata, Ayesha A. Motala, Katherine Ogurtsova, Jonathan E. Shaw, Dominic Bright, and Rhys Williams. 2019. Global and regional diabetes prevalence estimates for 2019 and projections for 2030 and 2045: Results from the International Diabetes Federation Diabetes Atlas. *Diabetes Research and Clinical Practice*, 157:107843.

Isabelle Scholl, Jördis M. Zill, Martin Härter, and Jörg Dirmaier. 2014. An Integrative Model of Patient-Centeredness – A Systematic Review and Concept Analysis. *PLOS ONE*, 9(9):e107828.

Clive Seale, Jonathan Charteris-Black, Aidan MacFarlane, and Ann McPherson. 2010. Interviews and Internet Forums: A Comparison of Two Sources of Qualitative Data. *Qualitative Health Research*, 20(5):595–606.

Marina Sokolova and Victoria Bobicev. 2013. What Sentiments Can Be Found in Medical Forums? In *Proceedings of Recent Advances in Natural Language Processing*, pages 633–639. Shoumen, Bulgaria: INCOMA Ltd.

Pontus Stenetorp, Sampo Pyysalo, Goran Topić, Tomoko Ohta, Sophia Ananiadou, and Jun'ichi Tsujii. 2012. brat: a Web-based Tool for NLP-Assisted Text Annotation. In *Proceedings of the Demonstrations at the 13th Conference of the European Chapter of the Association for Computational Linguistics*, pages 102–107. Association for Computational Linguistics.

Fabian Sudau, Tim Friede, Jens Grabowski, Janka Koschack, Philip Makedonski, and Wolfgang Himmel. 2014. Sources of Information and Behavioral Patterns in Online Health Forums: Observational Study. *Journal of Medical Internet Research*, 16(1):e10.

Martijn van der Eijk, Marjan J. Faber, Johanna W. M. Aarts, Jan A. M. Kremer, Marten Munneke, and Bastiaan R. Bloem. 2013. Using Online Health Communities to Deliver Patient-Centered Care to People With Chronic Conditions. *Journal of Medical Internet Research*, 15(6):e115.

Courtland VanDam, Shaheen Kanthawala, Wanda Pratt, Joyce Chai, and Jina Huh. 2017. Detecting clinically related content in online patient posts. *Journal of Biomedical Informatics*, 75:96–106.

Thomas D. Wilson. 2000. Human Information Behavior. *Informing Science*, 3(2):49–56.

Overview of the Fifth Social Media Mining for Health Applications (#SMM4H) Shared Tasks at COLING 2020

Ari Z. Klein
University of Pennsylvania
Philadelphia, PA, USA

Ilseyar Alimova
Kazan Federal University
Kazan, Russia

Ivan Flores
University of Pennsylvania
Philadelphia, PA, USA

Arjun Magge
University of Pennsylvania
Philadelphia, PA, USA

Zulfat Miftahutdinov
Kazan Federal University
Kazan, Russia

Anne-Lyse Minard
University of Orléans, LLL-CNRS
Orléans, France

Karen O'Connor
University of Pennsylvania
Philadelphia, PA, USA

Abeed Sarker
Emory University
Atlanta, GA, USA

Elena Tutubalina
Kazan Federal University
Kazan, Russia

Davy Weissenbacher
University of Pennsylvania
Philadelphia, PA, USA

Graciela Gonzalez-Hernandez
University of Pennsylvania
Philadelphia, PA, USA

{ariklein, ivan.flores, arjun.magge, karoc, dweissen, gragon}@pennmedicine.upenn.edu
{alimovailseyar, zulfatmi, tutubalinaev}@gmail.com
anne-lyse.minard@univ-orleans.fr, abeed@dbmi.emory.edu

Abstract

The vast amount of data on social media presents significant opportunities and challenges for utilizing it as a resource for health informatics. The fifth iteration of the Social Media Mining for Health Applications (#SMM4H) shared tasks sought to advance the use of Twitter data (tweets) for pharmacovigilance, toxicovigilance, and epidemiology of birth defects. In addition to re-runs of three tasks, #SMM4H 2020 included new tasks for detecting adverse effects of medications in French and Russian tweets, characterizing chatter related to prescription medication abuse, and detecting self reports of birth defect pregnancy outcomes. The five tasks required methods for binary classification, multi-class classification, and named entity recognition (NER). With 29 teams and a total of 130 system submissions, participation in the #SMM4H shared tasks continues to grow.

1 Introduction

The aim of the Social Media Mining for Health Applications (#SMM4H) shared tasks is to take a community-driven approach to addressing natural language processing (NLP) challenges of utilizing social media data for health informatics, including informal, colloquial expressions of clinical concepts, noise, data sparsity, ambiguity, and multilingual posts. The fifth iteration of the #SMM4H shared tasks consisted of five tasks involving mining health-related information from Twitter data (tweets): automatic classification of tweets that mention medications (Task 1), automatic classification of multilingual tweets that report adverse effects of a medication (Task 2), with sub-tasks for distinct sets of tweets posted in English (Task 2a), French (Task 2b), and Russian (Task 2c), automatic extraction and normalization of adverse effects in English tweets (Task 3), automatic characterization of chatter related to prescription medication abuse in tweets (Task 4), and automatic classification of tweets self-reporting a birth defect pregnancy outcome (Task 5).

Teams could register for one or multiple tasks. In total, 57 teams registered for at least one task. To develop their systems, teams were provided with annotated training and validation sets of tweets for

Proceedings of the 5th Social Media Mining for Health Applications (#SMM4H) Workshop & Shared Task, pages 27–36
Barcelona, Spain (Online), December 12, 2020.

each task. For the final evaluation, teams were provided with an unlabeled test set for each task, and were allowed up to four days to submit the predictions of their systems to CodaLab[1]—a platform that facilitates data science competitions. Each team was allowed to submit up to three sets of predictions per task. In total, 29 of the 57 registered teams submitted at least one set of predictions. More specifically, 16 teams participated in Task 1 (40 submissions), 17 teams in Task 2a (35 submissions), 5 teams in Task 2b (7 submissions), 7 teams in Task 2c (14 submissions), 7 teams in Task 3 (15 submissions), 3 teams in Task 4 (9 submissions), and 4 teams in Task 5 (10 submissions). In Section 2, we will briefly describe the tasks. In Section 3, we will present the performance and a brief summary of each team's best-performing system for each task. Appendix A provides the system description papers corresponding to the team numbers used in Section 3.

2   Tasks

2.1   Task 1: Automatic Classification of Tweets that Mention Medications

Task 1 is a binary classification task that involves distinguishing tweets that mention a medication or dietary supplement (annotated as "1") from those that do not (annotated as "0"). For this task, we used the definition of drug products and dietary supplements provided by the FDA (U.S. Food and Drug Administration, 2017). For the #SMM4H 2018 shared tasks (Weissenbacher et al., 2018), a data set was used that contained an artificially balanced distribution of the two classes. For #SMM4H 2020, the data set represents their natural, highly imbalanced distribution (Weissenbacher et al., 2019). Evaluating classifiers on this data set models more closely the detection of tweets that mention medications in practice. The training set contains 69,272 tweets, with only 181 (0.3%) tweets that mention a medication. The 9622 training tweets from #SMM4H 2018 were also provided, with 4975 tweets that mention a medication. The test set contains 29,687 tweets, with only 77 (0.3%) tweets that mention a medication. Systems were evaluated based on the F_1-score for the "positive" class (i.e., tweets that mention a medication).

2.2   Task 2: Automatic Classification of Multilingual Tweets that Report Adverse Effects

Task 2 is a binary classification task that involves distinguishing tweets that report an adverse effect of a medication (annotated as "1") from those that do not (annotated as "0"), with three sub-tasks for distinct sets of tweets posted in English, French, and Russian. The training set for the long-running, English-language version of this #SMM4H shared task contains 25,678 tweets, with 2377 (9.3%) tweets that report an adverse effect of a medication. The test set contains 4759 tweets, with 194 (4.1%) tweets that report an adverse effect.

For the French sub-task, the training set contains 2426 tweets, with only 39 (1.6%) tweets that report an adverse effect. The test set contains 607 tweets, with only 10 (1.6%) tweets that report an adverse effect. Inter-annotator agreement, based on dual annotations of 848 tweets by three annotators, was 0.61 and 0.69, for each of the two pairs of annotators.

For the Russian sub-task, the training set contains 7612 tweets, with 666 (8.7%) tweets that report an adverse effect. The test set contains 1903 tweets, with 166 (8.7%) tweets that report an adverse effect. All of the Russian tweets were dual annotated; first, three *Yandex.Toloka*[2] annotators' crowd-sourced labels were aggregated into a single label (Dawid and Skene, 1979), and then the tweets were labeled by a second annotator. Inter-annotator agreement was 0.49 (Cohen's kappa). Systems were evaluated based on the F_1-score for the "positive" class (i.e., tweets that report an adverse effect).

2.3   Task 3: Automatic Extraction and Normalization of Adverse Effects in English Tweets

Task 3 is a named entity recognition (NER) and entity normalization task that involves detecting the span of text within a tweet that reports an adverse effect of a medication, and normalizing the adverse effect to a unique Medical Dictionary for Regulatory Activities (MedDRA)[3] version 21.1 preferred term (PT) ID. The training set contains 2806 tweets, with 1829 (65%) tweets that report an adverse effect (annotated

---

[1] https://codalab.org/
[2] https://toloka.yandex.ru/
[3] https://www.meddra.org/

as "ADR"). For each tweet in the training set that reports an adverse effect, the span of text containing the adverse effect, the character offsets of that span of text, and the MedDRA ID of the adverse effect. The test set contains 1156 tweets, with 970 (84%) that report an adverse effect. Systems were evaluated based on their F_1-score, where a true positive is both the correct adverse effect (either partially or exactly matching the actual character offsets) and the correct MedDRA ID.

2.4 Task 4: Automatic Characterization of Prescription Medication Abuse Chatter in Tweets

Task 4 is a multi-class classification task that involves automatically distinguishing tweets mentioning potentially abuse-prone medications into one of four categories: (1) potential abuse/misuse (annotated as "A"), (2) non-abuse/misuse consumption (annotated as "C"), (3) medication mention only without any indication of consumption (annotated as "M"), and (4) unrelated (annotated as "U"). The medications mentioned in the tweets include prescription opioids, benzodiazepines, atypical anti-psychotics, central nervous system stimulants, and GABA (gamma aminobutyric acid) analogues. The training set contains 13,172 tweets: (1) 2133 (16%) "A" tweets, (2) 3668 (28%) "C" tweets, (3) 6843 (52%) "M" tweets, and (4) 528 (4%) "U" tweets. The test set contains 3271 tweets: (1) 503 (15%) "A" tweets, (2) 919 (28%) "C" tweets, (3) 1722 "M" (53%) tweets, and (4) 127 (4%) "U" tweets. Additional details about the data set, including the annotation process, annotation guidelines, and inter-annotator agreements, are presented in recent work (O'Connor et al., 2020). Systems were evaluated based on the F_1-score for the "potential misuse/abuse" ("A") class.

2.5 Task 5: Automatic Classification of Tweets Reporting a Birth Defect Pregnancy Outcome

Task 5 is a multi-class classification task that involves automatically distinguishing three classes of tweets that mention birth defects (Klein et al., 2018): (1) "defect" tweets refer to the user's child and indicate that he or she has the birth defect mentioned in the tweet (annotated as "1"); (2) "possible defect" tweets are ambiguous about whether someone is the user's child and/or has the birth defect mentioned in the tweet (annotated as "2"); (3) "non-defect" tweets merely mention birth defects (annotated as "3"). The training set contains 18,397 tweets: 966 (5%) "defect" tweets, 1041 (6%) "possible defect" tweets, and 16,390 (89%) "non-defect" tweets. The test set contains 4602 tweets: 244 (5%) "defect" tweets, 258 (6%) "possible defect" tweets, and 4100 (89%) "non-defect" tweets. Inter-annotator agreement, based on dual annotations for 21,727 of the tweets, was 0.86 (Cohen's kappa). Systems were evaluated based on the micro-averaged F_1-score for the "defect" and "possible defect" classes.

3 Results

3.1 Task 1: Automatic Classification of Tweets that Mention Medications

Table 1 presents the precision, recall, and F_1-score for the "positive" class (i.e., tweets that mention a medication), for each of the 16 team's best-performing system for Task 1. The majority of teams used a transformer-based architecture. Among these teams, the difference in performance seems to be based on the corpora used to pre-train the transformers, and the strategies used the address the high degree of class imbalance. The results suggest that imbalanced data remains a challenge for training deep neural network classifiers. The best-performing system for this task in #SMM4H 2018 (Weissenbacher et al., 2018) achieved an F_1-score of 0.918 (Chuhan et al., 2018) using an artificially balanced data set, while the best-performing system in #SMM4H 2020 achieved an F_1-score of 0.854. Nonetheless, advances in transformer-based architectures and strategies for addressing class imbalance have improved upon the baseline F_1-score of 0.788 (Weissenbacher et al., 2019).

3.2 Task 2: Automatic Classification of Multilingual Tweets that Report Adverse Effects

3.2.1 Automatic Classification of English Tweets that Report Adverse Effects

Table 2 presents the precision, recall, and F_1-score for the "positive" class (i.e., English tweets that report an adverse effect of a medication), for each of the 17 team's best-performing system for Task 2a. As in Task 1, the majority of teams used a transformer-based architecture. In particular, most of the better-

Team	F_1	P	R	System Summary
8	**0.85**	0.84	0.87	BERT, Bio+Clinical BERT, sub-corpus ensemble, SMM4H'18 corpus
3	0.80	0.77	0.83	RoBERTa pre-trained on tweets, ensemble, SMM4H'18 corpus
21	0.80	0.80	0.79	RoBERTa pre-trained on tweets
2	0.77	0.71	0.83	BERT, SMM4H'18 corpus
14	0.76	0.73	0.79	ELECTRA, decision tree, data augmentation, ensemble, SMM4H'18 corpus
17	0.76	0.82	0.70	BERT, DrugBank
18	0.76	0.77	0.74	RoBERTa pre-trained on biomedical literature
12	0.74	0.66	0.83	RoBERTa pre-trained on biomedical literature, over-sampling, SMM4H'18 corpus
23	0.72	0.84	0.62	BERT, BiLSTM, SMM4H'18 corpus
19	0.71	0.79	0.64	BioBERT pre-trained on tweets, sub-corpus ensemble, class weights, SMM4H'18 corpus
28	0.66	0.75	0.58	NA
9	0.64	0.74	0.56	decision tree, word and character 15-grams
29	0.60	0.57	0.64	NA
6	0.56	**0.86**	0.42	BioBERT, data augmentation, ensemble
22	0.45	0.57	0.38	NA
16	0.05	0.02	**0.90**	SVM, sent2vec sentence and bi-gram embeddings pre-trained on tweets, under-sampling

Table 1: Task 1 system summaries and F_1-scores (F_1), precision (P), and recall (R) for the "positive" class (i.e., tweets mentioning medications).

Team	F_1	P	R	System Summary
21	**0.64**	0.62	0.65	RoBERTa
25	0.58	**0.63**	0.54	EnDR-BERT, ensemble
10	0.58	0.52	0.65	RoBERTa, SMM4H'17 and SMM4H'19 corpora
5	0.57	0.50	0.66	RoBERTa
4	0.56	0.50	0.63	RoBERTa
7	0.56	0.56	0.55	RoBERTa, sub-corpus ensemble, rules
17	0.55	0.47	0.65	BERT, DrugBank, MedlinePlus, TransE MeSH representations
6	0.54	0.49	0.60	BioBERT, data augmentation, ensemble
1	0.51	0.48	0.54	CLAPA, BERT
22	0.48	0.44	0.53	SBERT RoBERTa sentence embeddings, class weights
2	0.47	0.58	0.40	BERT, SMM4H'20 Task 3 corpus
19	0.37	0.26	0.60	BioBERT pre-trained on tweets
20	0.35	0.28	0.46	CNN, GloVe word embeddings pre-trained on tweets, under-sampling
16	0.32	0.19	**0.87**	SVM, sent2vec sentence and bi-gram embeddings pre-trained on tweets, under-sampling
28	0.31	0.23	0.51	NA
15	0.31	0.31	0.31	logistic regression, feature engineering
29	0.27	0.16	0.79	NA

Table 2: Task 2a (English) system summaries and F_1-scores (F_1), precision (P), and recall (R) for the "positive" class (i.e., tweets reporting an adverse effect of a medication).

performing systems used RoBERTa-based models (Liu et al., 2019), with the best-performing system achieving an F_1-score of 0.64.

3.2.2 Automatic Classification of French Tweets that Report Adverse Effects

Table 3 presents the precision, recall, and F_1-score for the "positive" class (i.e., French tweets that report an adverse effect of a medication), for each of the five team's best-performing system for Task 2b. The highest F_1-score for the French-language version of this task is considerably lower than the highest F_1-scores for the automatic classification of adverse effects in English (0.64) and Russian (0.51) tweets. The difficulty of this task is further underscored by the fact that two teams were not able to detect any tweets reporting an adverse effect. This difficulty may be due to the small size of the training data and the high degree of class imbalance. To address the imbalanced data, Team 22 used a Bayesian optimization approach to class weighting, and Team 16 used under-sampling of the majority class.

3.2.3 Automatic Classification of Russian Tweets that Report Adverse Effects

Table 4 presents the precision, recall, and F_1-score for the "positive" class (i.e., Russian tweets that report an adverse effect of a medication), for each of the seven team's best-performing system for Task 2c. Teams 26 and 25 achieve the highest F_1-scores (0.51). Both teams used ensembles of BERT-based Russian language models from the DeepPavlov library (Burtsev et al., 2018). In addition, both teams used manually annotated drug reviews from the RuDREC corpus (Tutubalina et al., 2020) as additional

Team	F_1	P	R	System Summary
22	**0.17**	0.15	0.20	SBERT DistilBERT sentence embeddings, class weights
15	0.15	**0.33**	0.10	logistic regression, feature engineering
16	0.07	0.04	**0.60**	tree-based ensemble, LASER sentence embeddings, under-sampling
4	0.00	0.00	0.00	camemBERT
29	0.00	0.00	0.00	NA

Table 3: Task 2b (French) system summaries and F_1-scores (F_1), precision (P), and recall (R) for the "positive" class (i.e., tweets reporting an adverse effect of a medication).

training data, and Team 25 also used English drug reviews from the PsyTAR corpus ((Zolnoori et al., 2019).

Team	F_1	P	R	System Summary
26	**0.51**	0.45	0.60	Conversational RuBERT, under-sampling, ensemble, RuDReC corpus
25	0.51	**0.54**	0.48	EnRuDR-BERT, ensemble, bilingual training, RuDReC and PsyTAR corpora
5	0.48	0.36	0.70	RuBERT
22	0.42	0.35	0.55	SBERT DistilBERT sentence embeddings, class weights
4	0.36	0.34	0.40	RuBERT
28	0.36	0.29	0.46	NA
16	0.35	0.22	**0.89**	SVM, LASER sentence embeddings, under-sampling

Table 4: Task 2c (Russian) system summaries and F_1-scores (F_1), precision (P), and recall (R) for the "positive" class (i.e., tweets reporting an adverse effect of a medication).

3.3 Task 3: Automatic Extraction and Normalization of Adverse Effects in English Tweets

Table 5 presents the F_1-scores for the NER-based extraction of adverse effect text spans, and the precision, recall, and F_1-scores for the normalization to the MedDRA ID, for each of the seven team's best-performing systems for Task 3. Team 25 outperformed the other teams for all the presented performance metrics. For the NER-based extraction, they used a transformer-based architecture with domain-specific models, dictionary-based features, and additional training data from the CADEC corpus (Karimi et al., 2015). For normalization, they used a domain-specific, BERT-based classifier, additional training data, and similarity metrics comparing BERT-based word embeddings of Unified Medical Language System (UMLS) concepts and extracted NERs. Several other teams used similar approaches, so the performance of Team 26 might be attributed to their language models that were pre-trained specifically for detecting adverse drug reactions.

Team	E F_1	N F_1	N P	N R	System Summary
25	**0.76**	**0.46**	**0.48**	**0.45**	EnDR-BERT, dictionary, BERT-based similarity metrics, CADEC
2	0.73	0.38	0.34	0.44	BERT, CADEC, SMM4H'17 corpus
10	0.69	0.35	0.33	0.38	RoBERTa, multi-task learning
4	0.58	0.22	0.24	0.20	SciBERT/BioBERT/BERT ensemble, fastText-based similarity metrics, CADEC
1	0.46	0.20	0.35	0.14	BiLSTM, CRF, GloVe and EXT word embeddings, QuickUMLS
27	0.56	0.15	0.15	0.14	NA
16	0.16	0.00	0.00	0.00	dictionary

Table 5: Task 3 system summaries, F_1-scores (F_1) for adverse effect extraction (E), and F_1-scores (F_1), precision (P), and recall (R) for adverse effect normalization (N).

3.4 Task 4: Automatic Characterization of Prescription Medication Abuse Chatter in Tweets

Table 6 presents the precision, recall, and F_1-scores for the "potential abuse/misuse" class, for each team's best-performing system for Task 4. Team 13 achieved the highest F_1-score (0.51) using a CNN, fastText word embeddings, and data augmentation by means of manufacturing tweets that are semantically similar to the training data. This F_1-score, however, is lower than the F_1-score (0.67) of a stacked ensemble of BERT (Devlin et al., 2019), ALBERT (Lan et al., 2020), and RoBERTa models, presented in recent work (Ali Al-Garadi et al., 2020).

Team	F$_1$	P	R	System Summary
13	**0.51**	**0.53**	0.50	CNN, fastText word embeddings, data augmentation
1	0.49	0.46	0.51	SVM, under-sampling
16	0.46	0.35	**0.67**	SVM, sent2vec sentence and bi-gram embeddings pre-trained on tweets, under-sampling

Table 6: Task 4 system summaries and F$_1$-scores (F$_1$), precision (P), and recall (R) for the "potential abuse/misuse" class.

3.5 Task 5: Automatic Classification of Tweets Reporting a Birth Defect Pregnancy Outcome

Table 7 presents the micro-averaged precision, recall, and F$_1$-score for the "defect" and "possible defect" classes, for each team's best-performing system. Teams 6 and 24 achieved the highest micro-averaged F$_1$-scores (0.69). While Team 6 achieved a higher micro-averaged recall (0.73) than Team 24 (0.67) using a hard-voting ensemble of nine BioBERT-based models, Team 24 achieved a higher micro-averaged precision (0.71) than Team 6 (0.65) using ELMo word embeddings and data-specific resources for modeling birth defects, pregnancy-related information, people's names, and family relations. Team 19 also achieved a higher micro-averaged recall (0.69) than Team 24 (0.67) using BioBERT, but achieved a substantially lower micro-averaged precision (0.56) than Team 24 (0.71). Overall, for this imbalanced data, models based on contextualized word representations—BioBERT (Lee et al., 2020a) or ELMo (Peters et al., 2018)—outperformed a CNN-BiGRU neural network with GloVe word embeddings (Pennington et al., 2014). Recent work (Klein et al., 2019) presents baseline F$_1$-scores of an SVM classifier for the "defect" (0.65) and "possible defect" (0.51) classes.

Team	F$_1$	P	R	System Summary
6	**0.69**	0.65	**0.73**	BioBERT, data augmentation, ensemble
24	0.69	**0.71**	0.67	ELMo, GCNN, ANNIE NER, medical and family relations lexicons
19	0.62	0.56	0.69	BioBERT pre-trained on tweets
11	0.58	0.54	0.64	GloVe word and hashtag embeddings pre-trained on tweets, CNN, BiGRU

Table 7: Task 5 system summaries and micro-averaged F$_1$-score (F$_1$), precision (P), and recall (R) for the "defect" and "possible defect" classes.

4 Conclusion

This paper presented an overview of the #SMM4H 2020 shared tasks. With 29 teams and a total of 130 system submissions, participation in the #SMM4H shared tasks continues to grow. All of the teams with the best-performing system for each task used deep learning-based systems, most of which were transformer-based architectures. The system description papers that are cited in Appendix A were each peer-reviewed by two reviewers and provide further details about 26 teams' systems.

Acknowledgements

The work for #SMM4H 2020 at the University of Pennsylvania was supported by the National Institutes of Health (NIH) National Library of Medicine (NLM) [grant number R01LM011176]. The work at Kazan Federal University was supported by the Russian Science Foundation [grant number 18-11-00284]. The authors would also like to thank Alexis Upshur for her contribution to annotating tweets, Dmitry Ustalov and other members of the *Yandex.Toloka* team for providing credits for the crowd-sourced annotation of Russian tweets, and all those who reviewed system description papers.

References

Olanrewaju Tahir Aduragba, Jialin Yu, Gautham Senthilnathan, and Alexandra Cristea. 2020. Sentence contextual encoder with BERT and BiLSTM for automatic classification with imbalanced medication tweets. In *Proceedings of the Fifth Social Media Mining for Health Applications (#SMM4H) Workshop & Shared Task*, pages 165–167.

Mohammed Ali Al-Garadi, Yuan-Chi Yang, Haitao Cai, Yucheng Ruan, Karen O'Connor, Graciela Gonzalez-Hernandez, Jeanmarie Perrone, and Abeed Sarker. 2020. Text classification models for the automatic detection of nonmedical prescription medication use from social media. *medRxiv*.

Yandrapati Prakash Babu and Rajagopal Eswari. 2020. Identification of medication tweets using domain-specific pre-trained language models. In *Proceedings of the Fifth Social Media Mining for Health Applications (#SMM4H) Workshop & Shared Task*, pages 128–130.

Parsa Bagherzadeh and Sabine Bergler. 2020. CLaC at SMM4H 2020: Birth defect mention detection. In *Proceedings of the Fifth Social Media Mining for Health Applications (#SMM4H) Workshop & Shared Task*, pages 168–170.

Yang Bai and Xiaobing Zhou. 2020. Automatic detecting for health-related Twitter data with BioBERT. In *Proceedings of the Fifth Social Media Mining for Health Applications (#SMM4H) Workshop & Shared Task*, pages 63–69.

Pavel Blinov and Manvel Avetisian. 2020. Transformer models for drug adverse effects detection from tweets. In *Proceedings of the Fifth Social Media Mining for Health Applications (#SMM4H) Workshop & Shared Task*, pages 110–112.

Mikhail Burtsev, Alexander Seliverstov, Rafael Airapetyan, Mikhail Arkhipov, Dilyara Baymurzina, Nickolay Bushkov, Olga Gureenkova, Taras Khakhulin, Yurii Kuratov, Denis Kuznetsov, et al. 2018. Deeppavlov: Open-source library for dialogue systems. In *Proceedings of ACL 2018, System Demonstrations*, pages 122–127.

Silvia Casola and Alberto Lavelli. 2020. FBK@SMM4H2020: RoBERTa for detecting medications on Twitter. In *Proceedings of the Fifth Social Media Mining for Health Applications (#SMM4H) Workshop & Shared Task*, pages 101–103.

Wu Chuhan, Wu Fangzhao, Liu Junxin, Wu Sixing, Huang Yongfeng, and Xie Xing. 2018. Detecting tweets mentioning drug name and adverse drug reaction with hierarchical tweet representation and multi-head self-attention. In *Proceedings of the 2018 EMNLP Workshop SMM4H: The 3rd Social Media Mining for Health Applications Workshop and Shared Task*, pages 34–37.

Huong N. Dang, Kahyun Lee, Sam Henry, and Özlem Uzuner. 2020. Ensemble BERT for classifying medication-mentioning tweets. In *Proceedings of the Fifth Social Media Mining for Health Applications (#SMM4H) Workshop & Shared Task*, pages 37–41.

Alexander Philip Dawid and Allan M. Skene. 1979. Maximum likelihood estimation of observer error-rates using the EM algorithm. *Journal of the Royal Statistical Society: Series C (Applied Statistics)*, 28(1):20–28.

Jacob Devlin, Ming-Wei Chang, Kenton Lee, and Kristina Toutanova. 2019. BERT: Pre-training of deep bidirectional transformers for language understanding. In *Proceedings of the 17th Annual Conference of the North American Chapter of the Association for Computational Linguistics: Human Language Technologies (NAACL-HLT)*, pages 4171–4186.

George-Andrei Dima, Andrei-Marius Avram, and Dumitru-Clementin Cercel. 2020. Approaching SMM4H 2020 with ensembles of BERT flavours. In *Proceedings of the Fifth Social Media Mining for Health Applications (#SMM4H) Workshop & Shared Task*, pages 153–157.

Lucie Gattepaille. 2020. How far can we go with just out-of-the-box BERT models? In *Proceedings of the Fifth Social Media Mining for Health Applications (#SMM4H) Workshop & Shared Task*, pages 95–100.

Oguzhan Gencoglu. 2020. Sentence transformers and Bayesian optimization for adverse drug effect detection from Twitter. In *Proceedings of the Fifth Social Media Mining for Health Applications (#SMM4H) Workshop & Shared Task*, pages 161–164.

Andrey Gusev, Anna Kuznetsova, Anna Polyanskaya, and Egor Yatsishin. 2020. BERT implementation for detecting adverse drug effects mentions in Russian. In *Proceedings of the Fifth Social Media Mining for Health Applications (#SMM4H) Workshop & Shared Task*, pages 46–50.

Katikapalli Subramanyam Kalyan and Sivanesan Sangeetha. 2020. Want to identify, extract and normalize adverse drug reactions in tweets? Use RoBERTa. In *Proceedings of the Fifth Social Media Mining for Health Applications (#SMM4H) Workshop & Shared Task*, pages 121–124.

Sarvnaz Karimi, Alejanrdo Metke-Himenez, Madonna Kemp, and Chen Wang. 2015. Cadec: A corpus of adverse drug effect annotations. *Journal of Biomedical Informatics*, 55:73–81.

Sedigheh Khademi, Pari Delir Haghighi, and Frada Burstein. 2020. Adverse drug reaction detection in Twitter using RoBERTa and rules. In *Proceedings of the Fifth Social Media Mining for Health Applications (#SMM4H) Workshop & Shared Task*, pages 113–117.

Ari Z. Klein, Abeed Sarker, Haitao Cai, Davy Weissenbacher, and Graciela Gonzalez-Hernandez. 2018. Social media mining for birth defects research: A rule-based, bootstrapping approach to collecting data for rare health-related events on Twitter. *Journal of Biomedical Informatics*, 87:68–78.

Ari Z. Klein, Abeed Sarker, Davy Weissenbacher, and Graciela Gonzalez-Hernandez. 2019. Towards scaling Twitter data for digital epidemiology of birth defects. *npj Digital Medicine*, 2:1–9.

Zhenzhong Lan, Mingda Chen, Sebastian Goodman, Kevin Gimpel, Piyush Sharma, and Radu Soricut. 2020. ALBERT: A lite BERT for self-supervised learning of language representations. In *Proceedings of the 8th International Conference on Learning Representations (ICLR)*.

Jinhyuk Lee, Wonjin Yoon, Sundong Kim, Donghyeon Kim, Sunkyu Kim, Chan Ho So, and Jaewoo Kang. 2020a. BioBERT: A pretrained biomedical language representation model for biomedical text mining. *Bioinformatics*, 36(4):1234–1240.

Lung-Hao Lee, Po-Han Chen, Hao-Chuan Kao, Ting-Chun Hung, Po-Lei Lee, and Kuo-Kai Shyu. 2020b. Medication mention detection in tweets using ELECTRA transformers and decision trees. In *Proceedings of the Fifth Social Media Mining for Health Applications (#SMM4H) Workshop & Shared Task*, pages 131–133.

Mohamed Lichouri and Mourad Abbas. 2020. SpeechTrans@SMM4H'20: Impact of preprocessing and n-grams on automatic classification of tweets that mention medications. In *Proceedings of the Fifth Social Media Mining for Health Applications (#SMM4H) Workshop & Shared Task*, pages 118–120.

Yinhan Liu, Myle Ott, Naman Goyal, Jingfei Du, Mandar Joshi, Danqi Chen, Omer Levy, Mike Lewis, Luke Zettlemoyer, and Veselin Stoyanov. 2019. RoBERTa: A robustly optimized BERT pretraining approach. *arXiv Preprint*, arXiv:1907.11692.

Farhana Ferdousi Liza. 2020. Sentence classification with imbalanced data for health applications. In *Proceedings of the Fifth Social Media Mining for Health Applications (#SMM4H) Workshop & Shared Task*, pages 138–145.

Darshini Mahendran, Cora Lewis, and Bridget T. McInnes. 2020. NLP@VCU: Identifying adverse effects in English tweets for unbalanced data. In *Proceedings of the Fifth Social Media Mining for Health Applications (#SMM4H) Workshop & Shared Task*, pages 158–160.

Laiba Mehnaz. 2020. Automatic classification of tweets mentioning a medication using pre-trained sentence encoders. In *Proceedings of the Fifth Social Media Mining for Health Applications (#SMM4H) Workshop & Shared Task*, pages 150–152.

Isabel Metzger, Emir Y. Haskovic, Allison Black, Whitley M. Yi, Rajat S. Chandra1, Mark T. Rutledge, William McMahon, and Yindalon Aphinyanaphongs. 2020. SMM4H Shared Task 2020 - A hybrid pipeline for identifying prescription drug abuse from Twitter: Machine learning, deep learning, and post-processing. In *Proceedings of the Fifth Social Media Mining for Health Applications (#SMM4H) Workshop & Shared Task*, pages 57–62.

Zulfat Miftahutdinova, Andrey Sakhovskiy, and Elena Tutubalina. 2020. KFU NLP Team at SMM4H 2020 Tasks: Cross-lingual transfer learning with pretrained language models for drug reactions. In *Proceedings of the Fifth Social Media Mining for Health Applications (#SMM4H) Workshop & Shared Task*, pages 51–56.

Karen O'Connor, Abeed Sarker, Jeanmarie Perrone, and Graciela Gonzalez Hernandez. 2020. Promoting reproducible research for characterizing nonmedical use of medications through data annotation: Description of a Twitter corpus and guidelines. *Journal of Medical Internet Research*, 22(2):e15861.

Jeffrey Pennington, Richard Socher, and Christopher D. Manning. 2014. GloVe: Global vectors for word representation. In *Proceedings of the 2014 Conference on Empirical Methods in Natural Language Processing (EMNLP)*, pages 1532–1543.

Matthew E. Peters, Mark Neumann, Mohit Iyyer, and Matt Gardner. 2018. Deep contextualized word representations. In *Proceedings of the 16th Annual Conference of the North American Chapter of the Association for Computational Linguistics: Human Language Technologies (NAACL-HLT)*, pages 2227–2237.

Saichethan Miriyala Reddy. 2020. Detecting tweets reporting birth defect pregnancy outcome using two-view CNN RNN based architecture. In *Proceedings of the Fifth Social Media Mining for Health Applications (#SMM4H) Workshop & Shared Task*, pages 125–127.

Sougata Saha, Souvik Das, Prashi Khurana, and Rohini K. Srihari. 2020. Autobots ensemble: Identifying and extracting adverse drug reaction from tweets using transformer based pipelines. In *Proceedings of the Fifth Social Media Mining for Health Applications (#SMM4H) Workshop & Shared Task*, pages 104–109.

Ludovic Tanguy, Lydia-Mai Ho-Dac, Cécile Fabre, Roxane Bois, Touati Mohamed Yacine Haddad, Claire Ibarboure, Marie Joyau, François Le moal, Jade Moillic, Laura Roudaut, Mathilde Simounet, Irena Stankovic, and Mickaela Vandewaetere. 2020. LITL at SMM4H: An old-school feature-based classifier for identifying adverse effects in tweets. In *Proceedings of the Fifth Social Media Mining for Health Applications (#SMM4H) Workshop & Shared Task*, pages 134–137.

Elena Tutubalina, Ilseyar Alimova, Zulfat Miftahutdinov, Andrey Sakhovskiy, Valentin Malykh, and Sergey Nikolenko. 2020. The Russian Drug Reaction Corpus and neural models for drug reactions and effectiveness detection in user reviews. *Bioinformatics*.

U.S. Food and Drug Administration. 2017. Drugs@fda glossary of terms. `https://www.fda.gov/drugs/drug-approvals-and-databases/drugsfda-glossary-terms`. [Drug; Drug Product; online, accessed 21-July-2020].

V.G.Vinod Vydiswaran, Deahan Yu, Xinyan Zhao, Ermioni Carr, Jonathan Martindale, Jingcheng Xiao, Noha Ghannam, Matteo Althoen, Alexis Castellanos, Neel Patel, and Daniel Vasquez. 2020. Identifying medication abuse and adverse effects from tweets: University of Michigan at #SMM4H 2020. In *Proceedings of the Fifth Social Media Mining for Health Applications (#SMM4H) Workshop & Shared Task*, pages 90–94.

Chen-Kai Wang, You-Chen Zhang, Bo-Chun Xu, Bo-Hong Wang, You-Ning Xu, Po-Hao Chen, Hong-Jie Dai, and Chung-Hong Lee. 2020. ISLab system for SMM4H Shared Task 2020. In *Proceedings of the Fifth Social Media Mining for Health Applications (#SMM4H) Workshop & Shared Task*, pages 42–45.

Davy Weissenbacher, Abeed Sarker, Michael J. Paul, and Graciela Gonzalez-Hernandez. 2018. Overview of the third social media mining for health (SMM4H) shared tasks at EMNLP 2018. In *Proceedings of the 2018 EMNLP Workshop SMM4H: The 3rd Social Media Mining for Health Applications Workshop & Shared Task*, pages 13–16.

Davy Weissenbacher, Abeed Sarker, Ari Klein, Karen O'Connor, Arjun Magge, and Graciela Gonzalez-Hernandez. 2019. Deep neural networks for ensemble for detecting medication mentions in tweets. *Journal of the American Medical Informatics Association*, 26(12):1618–1626.

Xiaoyu Zhao, Ying Xiong, and Buzhou Tang. 2020. HITSZ-ICRC: A report for SMM4H shared task 2020-Automatic classification of medications and adverse effect in tweets . In *Proceedings of the Fifth Social Media Mining for Health Applications (#SMM4H) Workshop & Shared Task*, pages 146–149.

Maryam Zolnoori, Kin Wah Fung, Timothy B. Patrick, Paul Fontelo, Hadi Kharrazi, Anthony Faiola, Yi Shuan Shirley Wu, Christina E. Eldredge, Jake Luo, Mike Conway, et al. 2019. A systematic approach for developing a corpus of patient reported adverse drug events: A case study for SSRI and SNRI medications. *Journal of Biomedical informatics*, 90:103091.

Appendix A. Team Numbers and System Description Papers

Team	System Description Paper
1	(Vydiswaran et al., 2020)
2	(Gattepaille, 2020)
3	(Casola and Lavelli, 2020)
4	(Saha et al., 2020)
5	(Blinov and Avetisian, 2020)
6	(Bai and Zhou, 2020)
7	(Khademi et al., 2020)
8	(Dang et al., 2020)
9	(Lichouri and Abbas, 2020)
10	(Kalyan and Sangeetha, 2020)
11	(Reddy, 2020)
12	(Babu and Eswari, 2020)
13	(Metzger et al., 2020)
14	(Lee et al., 2020b)
15	(Tanguy et al., 2020)
16	(Liza, 2020)
17	(Zhao et al., 2020)
18	(Mehnaz, 2020)
19	(Dima et al., 2020)
20	(Mahendran et al., 2020)
21	(Wang et al., 2020)
22	(Gencoglu, 2020)
23	(Aduragba et al., 2020)
24	(Bagherzadeh and Bergler, 2020)
25	(Miftahutdinova et al., 2020)
26	(Gusev et al., 2020)
27	NA
28	NA
29	NA

Ensemble BERT for classifying medication-mentioning Tweets

Huong N. Dang, Kahyun Lee, Sam Henry, and Özlem Uzuner
George Mason University
Fairfax, VA, U.S.A.
{hdang20, klee70, shenry20, ouzuner}@gmu.edu

Abstract

Twitter is a valuable source of patient-generated data that has been used in various population health studies. The first step in many of these studies is to identify and capture Twitter messages (tweets) containing medication mentions. In this article, we describe our submission to Task 1 of the Social Media Mining for Health Applications (SMM4H) Shared Task 2020. This task challenged participants to detect tweets that mention medications or dietary supplements in a natural, highly imbalanced dataset. Our system combined a handcrafted preprocessing step with an ensemble of 20 BERT-based classifiers generated by dividing the training dataset into subsets using 10-fold cross validation and exploiting two BERT embedding models. Our system ranked first in this task, and improved the average F1 score across all participating teams by 19.07% with a precision, recall, and F1 on the test set of 83.75%, 87.01%, and 85.35% respectively.

1 Introduction

Social media platforms such as Twitter have been used in various public health studies. In these studies, detecting tweets containing health-related words such as diseases, treatments and medications is a fundamental yet difficult step. These difficulties are exacerbated by the short length and informal nature of tweets, which often contain non-standard grammar, frequent misspellings, many contractions, extensive slang, and combined symbols (emojis/emoticons) to express emotion. Despite these difficulties, unique insights can be gained from exploring Twitter data which motivates further research on this task.

Task 1 of SMM4H Shared Task 2020 (Klein et al., 2020) challenged participants to develop an automatic classification system to identify tweets mentioning medications or dietary supplements. The task was formulated as a binary classification task, in which given a set of tweets a system should predict the label for each tweet. Organizers created the UPennHLP Twitter Pregnancy Corpus (Weissenbacher et al., 2019a) and divided it into training, validation, and test sets. The training and validation datasets contained tweets labeled with either positive or negative labels indicating whether or not they contain a medication or dietary supplement mention. The test set contained tweets with labels removed, and was withheld from participants until a short evaluation period during which participants predicted labels for each tweet. Participants were evaluated based on their system's performance (F1 score for the positive class) at generating labels for the test set. The class distribution of the training, validation, and test sets was highly imbalanced. Only about 0.26% of the tweets mention medications, which is typical for this task and common in practice.

As a baseline the Shared Task organizers described the `Kusuri` classifier (Weissenbacher et al., 2019a). `Kusuri` was trained in a combination of the training and validation datasets and achieved good performance in detecting positive tweets in this test dataset at F1 score 78.79%. We sought to improve upon `Kusuri` by developing a system that combines a rule-based preprocessing component, a pre-filtering component, and an ensemble of BERT-based classifiers to predict the label of each tweet (Figure 1). In our best performing system, we used 10-fold cross validation to divide the data into 10

Proceedings of the 5th Social Media Mining for Health Applications (#SMM4H) Workshop & Shared Task, pages 37–41
Barcelona, Spain (Online), December 12, 2020.

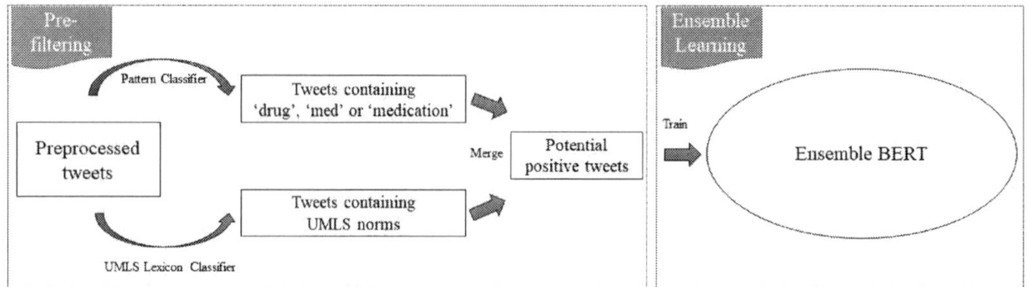

Figure 1: An overview of the system for automatic classification of tweets mentioning medications

subsets, then trained a *Bio+Clinical BERT* classifier and a *BERT-Large Uncased* classifier on each subset to obtain 20 classifiers. These 20 classifiers were ensembled using the average of their predicted probabilities to predict the final class labels. We further increased the performance of this system by training it on additional data from SMM4H 2018 which consisted of a very similar task.

2 Methods

2.1 Preprocessing

Preprocessing removes noise from tweets and converts words to a form that is recognizable by the embedding system. All tweets were pre-processed as follows: (1) Username, URL, Retweet abbreviation 'rt', non-ASCII words and characters were removed; (2) Artifacts and upstream processing such as '&', '<', '>' were standardized to 'and', '<', and '>' respectively; (3) Camel-cased expressions in hashtags were split into their component words and the hash symbols were removed; (4) A space was added between a word and a punctuation; (5) All characters were lowercased; (6) Contractions were converted to their formal forms.

2.2 Pre-filtering

Next, the preprocessed tweets were pre-filtered to remove some negative tweets. Pre-filtering was effective at removing approximately 25% of the negative tweets, and the class ratio increased from 0.26% positive class before pre-filtering to 0.34% positive class after pre-filtering. Pre-filtering consisted of a Pattern Classifier and a Unified Medical Language System Metathesaurus (UMLS) Lexicon Classifier. Tweets were labeled as *Potential positive tweets* if the requirements of either classifier were met. The Pattern Classifier labeled tweets containing the words 'drug', 'medication' or 'med' as *Potential positive tweets* based on the assumption that these tweets are likely to contain medication mentions. The UMLS Lexicon Classifier selected tweets containing at least one concept from sections 0, 1, 2 and 9 of the UMLS 2017 AA release. Given a tweet, the UMLS Lexicon classifier identified entities in that tweet and checked if those entities were linked to any concepts within a default threshold of 0.7. The UMLS Lexicon classifier was implemented using the Name Entity Recognition (NER) model and Entity Linking model of scispaCy *en_core_sci_lg* package (Neumann et al., 2019). The scispaCy NER model was trained from nine datasets covering a diversity of entities in several biomedical domains. The scispaCy Entity Linking model is capable of linking an entity to UMLS concepts and concept *aliases* which include common names of drugs and alternative spellings.

2.3 Ensemble Learning

Lastly, to predict whether each tweet contained a medication or dietary supplement mention, we used an ensemble of BERT-based linear classifiers. BERT (Devlin et al., 2018) is a pre-trained contextual word representation encoder based on transformers and can be fine-tuned using task-specific layers in addition to BERT for a specific task. BERT has been used in state-of-the-art systems for several Natural Language Processing (NLP) tasks, including winning systems of the SMM4H 2019 Shared Task (Weissenbacher

Model	Bio+Clinical BERT			BERT-Base			BERT-Large		
Method	**Single**	**10-fold ensemble**		**Single**	**10-fold ensemble**		**Single**	**10-fold ensemble**	
		Average	**Hard Voting**		**Average**	**Hard Voting**		**Average**	**Hard Voting**
F1	83.87	83.58	73.68	81.16	84.38	79.45	73.17	*84.85*	79.37
Precision	*96.30*	87.5	68.29	82.35	*93.10*	76.32	63.83	90.32	89.29
Recall	74.29	80.00	80.00	80.00	77.14	82.86	85.71	80.00	71.43

Table 1: Performance on SMM4H 2020 validation dataset of the single models and 10-fold ensemble models using *individual* BERT models

Model	Bio+Clinical BERT and BERT-Base	Bio+Clinical BERT and BERT-Large	BERT-Base and BERT-Large	Bio+Clinical BERT, BERT-Large and BERT-Base
F1	84.85	*86.15*	85.71	83.08
Precision	90.32	93.33	*96.43*	90.00
Recall	*80.00*	*80.00*	77.14	77.14

Table 2: Performance on SMM4H 2020 validation dataset of multiple *combined-BERT* models in addition to 10-fold ensemble

et al., 2019b) which focused on using machine learning techniques to detect tweets containing adverse drug reactions (Task 1) or personal health mentions (Task 4).

Since the tweets in this shared task belong to the consumer health domain, they contain language characteristic of the general English, clinical, and biomedical domains. We experimented with three different pre-trained BERT models: two from the general English domain, *BERT-Base Uncased* and *BERT-Large Uncased* (Devlin et al., 2018), and one from the biomedical and clinical domains, *Bio+Clinical BERT* (Alsentzer et al., 2019). *Bio+Clinical BERT* uses *Bio-BERT* (Lee et al., 2020) which is trained on biomedical text, to initialize embeddings which are then trained on clinical texts. It was shown to outperform *Bio-BERT* and *BERT-Base Uncased* on three of five common Clinical NLP tasks (Alsentzer et al., 2019). We fine-tuned each of these BERT models using a linear classification layer with AdamW optimizer for four epochs with a batch size of 16 and a learning rate of 0.00001 in a 10-fold cross-validation setup before starting the ensemble process.

For each pre-trained word embedding model, *BERT-Base Uncased*, *BERT-Large Uncased* and *Bio+Clinical BERT*, we trained both a single model and 10-fold-ensemble model. In the single model, the training dataset was split into training and validating subsets with ratio 90:10, and the trained model was used to predict labels for tweets in the validation dataset. In the 10-fold-ensemble model, 10 different models were created from the data splits of 10-fold cross-validation. Therefore each separate model had its own training and validating subsets. We experimented with two ensemble methods to combine the outputs of these systems: (1) Average, in which we take the mean of predicted probabilities of each individual classifier and use argmax to obtain the class label, (2) Hard Voting, in which we select the majority class of the class labels predicted by each individual classifier.

To balance out the individual weaknesses of the three pre-trained BERT models, we also performed four more experiments which combined *Bio+Clinical BERT*, *BERT-Base Uncased*, and *BERT-Large Uncased* ensemble models into a single ensemble. We experimented by combining each pair, and all three models using the average ensemble method to combine their predictions.

2.4 Additional Training Data

After training our initial model, we performed an error analysis on the validation set to find reasons for false negative predictions. We found that although each false negative sample included a word or a phrase related to medications, their predicted positive probability was low. These samples typically contained either (1) medication-related words or phrases that were either unseen or rarely seen in the training dataset, or (2) contained medication-related words or phrases which were inconsistently labeled within the training dataset. To correct for these errors, we added data from the SMM4H Shared Task

Model	Bio+Clinical BERT	BERT-Base	BERT-Large	Bio-Clinical-BERT, BERT-Base	Bio-Clinical-BERT, BERT-Large	BERT-Base, BERT-Large	Bio-Clinical-BERT, BERT-Large, BERT-Base
F1	84.06	84.93	83.33	**89.55**	**89.55**	84.51	86.96
Precision	85.29	81.58	81.08	93.75	93.75	83.33	88.24
Recall	82.86	88.57	85.71	85.71	85.71	85.71	85.71

Table 3: Performance on SMM4H 2020 validation dataset of *individual* and *combined-BERT* models in addition to 10-fold ensemble using additional training data

2018 on classifying tweets mentioning drug names or dietary supplement (Weissenbacher et al., 2018) to the training set. This dataset contained labeled samples for most of the false-negatives predictions our system made and contained 4,975 positive and 4,647 negative tweets.

3 Results

3.1 System Development

Table 1 compares performance of the single models and 10-fold ensemble models on the SMM4H 2020 validation dataset using each pre-trained BERT model. For all 10-fold ensemble models, the F1 score of the average method was higher than that of the hard voting method, and the average method was either very similar to or higher than the single, non-ensemble model. Table 1 also shows that *BERT-Large Uncased* 10-fold-ensemble model achieved the highest F1 score at 84.85%, indicating it is the most efficient model using single BERT model. However, we found that combining multiple BERT models increased performance. Table 2 shows that when combining 10-fold ensemble models, the combination of *Bio+Clinical BERT* and *BERT-Large Uncased* performs the best with F1 score at 86.15%.

Tables 1 and 2 show performance when using only the SMM4H 2020 training data. Table 3 shows the performance of multiple average ensemble models using both the SMM4H 2020 and 2018 data for training. Both *Bio+Clinical BERT* and *BERT-Large Uncased* ensemble model and *Bio+Clinical BERT* and *BERT-Base Uncased* ensemble model achieved the highest F1 score at 89.55%. This performance increased by 3.4% compared to the best *Bio+Clinical BERT* and *BERT-Large Uncased* ensemble model without additional data in Table 2. This shows that including related training data improved performance.

3.2 System Submissions

We made three system submissions for test dataset evaluation. Each of these systems used the best configuration indicated by our validation set experiments, and consisted of an ensemble of 20 *BERT-Large Uncased* and *Bio+Clinical BERT* classifiers using the average ensemble method. Each system was trained on a combination of the training and validation datasets from SMM4H 2020, and the training dataset of SMM4H 2018. The submissions differed in their use of the pre-filtering phase. The first submission used pre-filtering for training and evaluation. The second submission did not use pre-filtering for training, but used it for evaluation. The third submission did not use pre-filtering at all. The third submission obtained the highest F1 score on the test dataset of Shared Task 2020 at 85.35%, followed by the second variant and the first variant with F1 score at 85% and 83% respectively. These results indicate that pre-filtering does not help the system classify tweets on the test dataset. The third submission is the highest performing system in Task 1 of the SMM4H Shared Task 2020. In addition, its F1 score outperformed that of the baseline system `Kusuri` by 6.56%.

4 Conclusion

In this work, we developed an ensemble tweet classifier based on BERT combined with pre-processing and pre-filtering steps. The system was evaluated on a natural, highly imbalanced dataset, achieved the highest F1 score in Task 1 of the SMM4H Shared Task 2020. We found that the best system configuration consisted of an ensemble of 20 BERT-based classifiers built using *BERT-Large Uncased* and *Bio+Clinical BERT* ensembled using an average method. We also found that including related training data improved performance, and that the pre-filtering step did not improve performance. For future work, we plan to experiment with different neural network classification layers in addition to BERT outputs.

References

Emily Alsentzer, John R Murphy, Willie Boag, Wei-Hung Weng, Di Jin, Tristan Naumann, and Matthew McDermott. 2019. Publicly available clinical BERT embeddings. *arXiv preprint arXiv:1904.03323*.

Jacob Devlin, Ming-Wei Chang, Kenton Lee, and Kristina Toutanova. 2018. Bert: Pre-training of deep bidirectional transformers for language understanding. *arXiv preprint arXiv:1810.04805*.

Ari Klein, Ilseyar Alimova, Ivan Flores, Arjun Magge, Zulfat Miftahutdinov, Anne-Lyse Minard, Karen O'Connor, Abeed Sarker, Elena Tutubalina, Davy Weissenbacher, and Graciela Gonzalez-Hernandez. 2020. Overview of the fifth Social Media Mining for Health Applications (#SMM4H) Shared Tasks at COLING 2020. In *Proceedings of the Fifth Social Media Mining for Health Applications (#SMM4H) Workshop & Shared Task*.

Jinhyuk Lee, Wonjin Yoon, Sungdong Kim, Donghyeon Kim, Sunkyu Kim, Chan Ho So, and Jaewoo Kang. 2020. BioBERT: a pre-trained biomedical language representation model for biomedical text mining. *Bioinformatics*, 36(4):1234–1240.

Mark Neumann, Daniel King, Iz Beltagy, and Waleed Ammar. 2019. ScispaCy: Fast and Robust Models for Biomedical Natural Language Processing. In *Proceedings of the 18th BioNLP Workshop and Shared Task*, pages 319–327, Florence, Italy, August. Association for Computational Linguistics.

Davy Weissenbacher, Abeed Sarker, Michael J. Paul, and Graciela Gonzalez-Hernandez. 2018. Overview of the third social media mining for health (SMM4H) shared tasks at EMNLP 2018. In *Proceedings of the 2018 EMNLP Workshop SMM4H: The 3rd Social Media Mining for Health Applications Workshop & Shared Task*, pages 13–16, Brussels, Belgium, October. Association for Computational Linguistics.

Davy Weissenbacher, Abeed Sarker, Ari Klein, Karen O'Connor, Arjun Magge, and Graciela Gonzalez-Hernandez. 2019a. Deep neural networks ensemble for detecting medication mentions in tweets. *Journal of the American Medical Informatics Association*, 26(12):1618–1626.

Davy Weissenbacher, Abeed Sarker, Arjun Magge, Ashlynn Daughton, Karen O'Connor, Michael J. Paul, and Graciela Gonzalez-Hernandez. 2019b. Overview of the fourth social media mining for health (SMM4H) shared tasks at ACL 2019. In *Proceedings of the Fourth Social Media Mining for Health Applications (#SMM4H) Workshop & Shared Task*, pages 21–30, Florence, Italy, August. Association for Computational Linguistics.

ISLab System for SMM4H Shared Task 2020

Chen-Kai Wang[1,2], You-Chen Zhang[3], Bo-Chun Xu[3], Bo-Hong Wang[3], You-Ning Xu[3], Po-Hao Chen[3], Hong-Jie Dai[3,4,5*], Chung-Hong Lee[3]

[1] Big Data Laboratories, Chunghwa Telecom Laboratories, Taoyuan, Taiwan, R.O.C
[2] Department of Computer Science, National Chiao Tung University, Hsinchu, Taiwan, R.O.C.
[3] Department of Electrical Engineering, College of Electrical Engineering and Computer Science, National Kaohsiung University of Science and Technology, Kaohsiung, Taiwan, R.O.C.
[4] Department of Post-Baccalaureate Medicine, College of Medicine, Kaohsiung Medical University, Kaohsiung, Taiwan, R.O.C.
[5] National Institute of Cancer Research, National Health Research Institutes, Tainan, Taiwan, R.O.C.

{dennisckwang, magicpower503, xobochun0116}@gmail.com, {1061247131, 108111109, f107154153, hjdai}@nkust.edu.tw, leechung@mail.ee.kuas.edu.tw

Abstract

In this paper, we described our systems for the first and second subtasks of Social Media Mining for Health Applications (SMM4H) shared task in 2020. The two subtasks are automatic classification of medication mentions and adverse effect in tweets. Our systems for both subtasks are based on Robustly optimized BERT approach (RoBERTa) and our previous work at SMM4H'19. The best F1-scores achieved by our systems for subtask 1 and 2 were 0.7974 and 0.64 respectively, which outperformed the average F1-scores among all teams' best runs by at least 0.13.

1 Introduction

Nowadays, social media are often being used by general public to create and share public messages related to their health. With the global increase in social media usage, there is a trend of posting information related to adverse drug reactions (ADR). Mining social media data for this type of information is helpful for pharmacological post-marketing surveillance and monitoring. In order to facilitate the use of social media for health monitoring and surveillance, we participated in the social media mining for health applications (SMM4H) shared task to develop systems that can automatically identify tweets conveying medications and adverse effects.

2 Task and Data Description

2.1 Task 1: Automatic Classification of Tweets that Mention Medications

This task is a binary classification task involves distinguishing tweets that determine whether it mentions medications or dietary supplements. The organizers provided a training set consisting of 69,272 tweets for all participants to develop their system, and a test set consisting of 29,687 tweets. Table 1 shows the distribution of the binary labels over the training and test sets. We can find that the training set is highly imbalanced.

Table 1: Distributions of labels over the training, validation and test datasets of task 1.

Dataset	Positive (1)	Negative (0)	Total
Training set	146 (0.26%)	55,273 (99.74%)	55,419
Validation set	35 (0.25%)	13,818 (99.75%)	13,853
Test set	N/A	N/A	29,687

Proceedings of the 5th Social Media Mining for Health Applications (#SMM4H) Workshop & Shared Task, pages 42–45
Barcelona, Spain (Online), December 12, 2020.

2.2   Task 2: Automatic Classification of Multilingual Tweets that Report Adverse Effects

Task 2 is also a binary classification task, which involves distinguishing tweets mentions ADRs. As illustrated in Table 2, the training set of the task is also highly imbalanced.

Table 2: Distribution of labels over the training, validation and test datasets of task 2.

Dataset	Positive (1)	Negative (0)	Total
Training set	1,903 (9.26%)	18,641 (90.74%)	20,544
Validation set	474 (9.23%)	4,660 (90.77%)	5,134
Test set	N/A	N/A	4,759

3   Methods

For each task, we developed three systems; the first and the second was based on Robustly Optimized BERT Pretraining Approach (RoBERTa) (Y. Liu et al., 2019) and the third was based on the method proposed in our previous work Dai and Wang (2019); (Wang et al., 2019).

3.1   System 1 and 2: RoBERTa and Retrained RoBERTA

BERT is an unsupervised language representation method to obtain deep bidirectional representations of sentences by jointly conditioning on both left and right context in all layers from free text. RoBERTa (Y. Liu et al., 2019) is an enhanced version of BERT which was trained with dynamic masking, full sentence without next sentence prediction loss, large mini-batches and a larger byte-level byte-pair encoding (BPE) (Sennrich et al., 2015). BPE is a hybrid between character- and word-level representations allowing handling large vocabularies in natural language corpora. Radford et al. (2019) introduced a clever implementation of BPE by using bytes instead of Unicode characters as the base sub-word units, which makes it possible to learn a sub-word vocabulary of a modest size that can still encode any input text without introducing any unknown tokens. In our implementation, we encoded both datasets released by the SMM4H organizers through BPE, and fine-tuned the RoBERTa-large architecture pre-trained model on the released training set to develop our first system.

For the system 2, we retrained the RoBERTa-large architecture on a tweet unlabeled corpus collected by our team. The unlabeled corpus consisted of 6,339,457 tweets on Twitter collected from January to April 2019, according to 183,593 drug names recorded in RxNorm (S. Liu et al., 2005) and 13,699 ADRs released by Nikfarjam et al. (2015).

For both models, we set the same parameters to fine-tuned RoBERTa; each model was trained by using the Adam optimizer with a learning rate of 10^{-5} for 10 epochs with a batch size of 8.

3.2   System 3: Random Under Sampling with Word Embedding-based Synthetic Minority Over-sampling Technique

Because the datasets of both subtasks are highly imbalanced, we applied the word embedding-based synthetic minority over-sampling technique (WESMOTE) with Random Under Sampling (RUS) proposed in our previous work (DaiandWang, 2019; Wang et al., 2019) to develop classifiers with reliable performance. In our implementation, we first applied WESMOTE to synthesize new positive examples by using the sentence representation based on BERT. We then randomly under-sampled the negative examples so that the ratio of positive against negative is 1:2. In order to extract features for training our classifiers, we pre-processed tweets to replace URLs, dosages and Twitter specific characters with the corresponding symbols, and modified the numeral parts in each token to one as proposed in our previous work (Dai et al., 2016). The preprocessed tweet was then processed by a tweet tokenizer (Owoputi et al., 2013) to generate tokens. Follow by the above step, each token was processed by Hunspell (Anonymous, 2019) to detect spelling errors. If a token is considered to be misspelled, the first recommended correction is included as an alternative term for the token. Finally, we lowercased all tokens and used the Snowball stemmer (Porter, 2001) to perform stemming without removing any stop words. After the above steps, we extracted the following features to train our support vector machine (SVM) model:

- Bag-of-word features: we extracted unigram and bigram with TF-IDF (Term Frequency-Inverse Document Frequency) as the weighting scheme.

- Domain knowledge features: The presence of adverse drug reaction (ADR) or drug mentions were engineered as two binary features with the value of either 0 or 1. The occurrences of ADR and drug names were recognized by using the ADR mention recognizer developed in our previous work (Dai et al., 2016; Wang et al., 2018).
- Negation features: The feature set uses three flags to indicate the occurrence of an ADR mention is missing, positive or negated. If a tweet contains ADRs, the NegEx algorithm (Chapman et al., 2001) is employed to determine whether the occurrence is negated.
- Word embedding features: The features were generated by taking the mean across all tokens' embedding represented by a 1024-dimensional vector based on the whole word masking variant of BERT-Large released by Turc et al. (2019).

4   Results

Table 3 and Table 4 show the performance of our systems for task 1 and task 2 on the validation and test data, respectively. The pre-trained RoBERTa model achieved the best F1-scores on both tasks. Compared with RUS_WESMOTE and Retrained RoBERTa on the two validation sets, RoBERTa exhibited much better performance too.

Table 3: Performance on validation and test data for task 1.

System	Validation			Test		
	P	R	F	P	R	F
RoBERTa	**0.938**	**0.857**	**0.896**	0.803	0.792	**0.797**
RUS_WESMOTE	0.484	0.534	0.508			0.47
Retrained RoBERTA	0.558	0.686	0.615			0.48
Average scores				0.7032	0.6948	0.6628

Table 4: Performance on validation and test data for task 2.

System	Validation			Test		
	P	R	F	P	R	F
RoBERTa	**0.662**	**0.711**	**0.686**	0.62	0.65	**0.64**
RUS_WESMOTE	0.544	0.527	0.535			0.41
Retrained RoBERTA	0.481	0.627	0.544			0.45
Average scores				0.42	0.59	0.46

5   Discussion

For both subtasks, the distribution of binary class is highly imbalanced. When we used the traditional machine learning methods like SVM, even we have tried to deal with the data imbalance problem by RUS_WESMOTE, the performance is still far below that of the system with the pre-trained RoBERTa. On the other hand, for the RoBERTa based systems, we didn't apply any imbalance techniques but they still get compatible and even better precision, recall and F-score. It is surprising to see that the system based on the tweet-retrained RoBERTa model didn't outperform the original pre-trained RoBERTa model.

6   Conclusion

We demonstrated that the system based on the pre-trained RoBERTa model outperformed traditional SVM-based method and the re-trained RoBERTa model. We will conduct error analysis to interpret the results of the two RoBERTa-based models to figure out the reason why the performance of the re-trained RoBERTa model get worse in the future.

Acknowledgements

This study was supported by the Ministry of Science and Technology of Taiwan [Grant numbers MOST-106-2221-E-143-007-MY3] and [Grant numbers MOST 109-2221-E-992-074-MY3].

Reference

Anonymous. (2019). Hunspell. Retrieved from http://hunspell.github.io/

Chapman, Wendy W, Bridewell, Will, Hanbury, Paul, Cooper, Gregory F, & Buchanan, Bruce G. (2001). A simple algorithm for identifying negated findings and diseases in discharge summaries. Journal of biomedical informatics, 34(5), 301-310.

Dai, Hong-Jie, Touray, Musa, Jonnagaddala, Jitendra, & Syed-Abdul, Shabbir. (2016). Feature engineering for recognizing adverse drug reactions from twitter posts. Information, 7(2), 27.

Dai, Hong-Jie, & Wang, Chen-Kai. (2019). Classifying adverse drug reactions from imbalanced Twitter data. International journal of medical informatics, 129, 122-132.

Liu, Simon, Ma, Wei, Moore, Robin, Ganesan, Vikraman, & Nelson, Stuart %J IT professional. (2005). RxNorm: prescription for electronic drug information exchange. 7(5), 17-23.

Liu, Yinhan, Ott, Myle, Goyal, Naman, Du, Jingfei, Joshi, Mandar, Chen, Danqi, . . . Stoyanov, Veselin. (2019). Roberta: A robustly optimized bert pretraining approach. arXiv preprint arXiv:1907.11692.

Nikfarjam, Azadeh, Sarker, Abeed, O'connor, Karen, Ginn, Rachel, & Gonzalez, Graciela %J Journal of the American Medical Informatics Association. (2015). Pharmacovigilance from social media: mining adverse drug reaction mentions using sequence labeling with word embedding cluster features. 22(3), 671-681.

Owoputi, Olutobi, O'Connor, Brendan, Dyer, Chris, Gimpel, Kevin, Schneider, Nathan, & Smith, Noah A. (2013). Improved part-of-speech tagging for online conversational text with word clusters. Paper presented at the Proceedings of the 2013 conference of the North American chapter of the association for computational linguistics: human language technologies.

Porter, Martin F. (2001). Snowball: A language for stemming algorithms, 2001. In.

Radford, Alec, Wu, Jeffrey, Child, Rewon, Luan, David, Amodei, Dario, & Sutskever, Ilya %J OpenAI Blog. (2019). Language models are unsupervised multitask learners. 1(8), 9.

Sennrich, Rico, Haddow, Barry, & Birch, Alexandra. (2015). Neural machine translation of rare words with subword units. arXiv preprint arXiv:1508.07909.

Turc, Iulia, Chang, Ming-Wei, Lee, Kenton, & Toutanova, Kristina. (2019). Well-read students learn better: On the importance of pre-training compact models. arXiv preprint arXiv:1908.08962.

Wang, Chen-Kai, Dai, Hong-Jie, & Wang, Bo-Hung. (2019). BIGODM System in the Social Media Mining for Health Applications Shared Task 2019. Paper presented at the Proceedings of the Fourth Social Media Mining for Health Applications (# SMM4H) Workshop & Shared Task.

Wang, Chen-Kai, Dai, Hong-Jie, Wang, Feng-Duo, & Su, Emily Chia-Yu. (2018). Adverse drug reaction post classification with imbalanced classification techniques. Paper presented at the 2018 Conference on Technologies and Applications of Artificial Intelligence (TAAI).

BERT implementation for detecting adverse drug effects mentions in Russian

Andrey Gusev[*]   **Anna Kuznetsova**[*]   **Anna Polyanskaya**[*]   **Egor Yatsishin**[*]

National Research University Higher School of Economics, Moscow

{aagusev_2,adkuznetsova_3,akpolyanskaya,esyatsishin}@edu.hse.ru

Abstract

This paper describes a system developed for the Social Media Mining for Health (SMM4H) 2020 shared task. Our team participated in the second subtask for Russian language creating a system to detect adverse drug reaction (ADR) presence in a text. For our submission, we exploited an ensemble model architecture, combining BERT's extension for Russian language, Logistic Regression and domain-specific preprocessing pipeline. Our system was ranked first among others, achieving F-score of 0.51. We have made our code publicly available[1].

1   Introduction

In this paper, we focus on the problem of discovering the presence of adverse drug reaction (ADR) concepts in twitter posts as part of the The Social Media Mining for Health Applications (SMM4H) Shared Task (Klein et al., 2020). The paper is based on the participation of our team in the Russian language segment of the second task: ADR presence classification. Organizers of SMM4H 2020 Task 2 provided datasets of Russian tweets with binary annotation indicating the presence or absence of ADRs in each post. The aim of the task was to develop a system to classify the tweets according to the presence of ADRs. Texts were given in a raw form, so they contained misspellings, slang, emojis, hashtags, usernames and were quite noisy. This year is the first time for a distinct set of Russian tweets to be included in the task. We tested and compared several different approaches for solving such type of classification task, including classical Machine Learning approaches and neural networks, and also different preprocessing pipelines.

2   Data

2.1   Datasets

The main dataset consists of the training set (7,612 tweets), validation set (1,522 tweets) and test set (1,903 tweets). The dataset is highly imbalanced with only 666 tweets labeled as mentioning ADR (hardly 9%). We used several techniques to overcome this problem, one of them being an attempt to create some additional data. We used RuDReC corpus[2] (Tutubalina et al., 2020) and manually labeled about 1,800 drug reviews (627 positive and 1195 negative).

It would be wrong to assume that these drug reviews are completely identical to the tweets from the main set in terms of linguistic features, so we did a simple analysis, which gave us several insights. First of all, the main lexicon of this two types of texts is quite similar, with the only two substantial differences:

1. Reviews tend to mention the cost or, to be more specific, the expensiveness of a drug much more often than tweets do, leading to the higher distribution of words like *expensive* and *overpriced*;

---

[1]https://github.com/toskn/nru_hse_team_hlpl2020

[2]https://github.com/cimm-kzn/RuDReC

Proceedings of the 5th Social Media Mining for Health Applications (#SMM4H) Workshop & Shared Task, pages 46–50
Barcelona, Spain (Online), December 12, 2020.

2. In general, reviewers' language is significantly more grammatically correct and uses less slang and word shortenings.

The results of using this extended dataset are described in Section 4.

2.2 Preprocessing

Our approach for text preprocessing is to some extent based on the one used in (Ellendorff et al., 2019). The following changes were made using Python programming language:

- Tokenization using CrazyTokenizer from RedditScore package[3] based on spaCy tokenizer for Russian[4];
- Lowercasing, except all-caps words such as АД "antidepressants";
- All ё replaced with е;
- Urls replaced with ЮРЛ "url" placeholder;
- Usernames replaced with ЮЗЕРНЕЙМ "username" placeholder;
- Hashtags replaced with ХЕШТЕГ "hashtag" placeholder;
- Numbers replaced with NUM placeholder;
- Measures such as кг "kg" and мл "ml" replaced with MEASURE placeholder;
- Emojis replaced with POS_EMOJI for ones with positive meanings, NEG_EMOJI for ones with negative meanings and NEUTRAL_EMOJI for ones related to health issues as they could be semantically important;
- 3 or more repetitive letters normalized to 1;
- Line breaks deleted;
- Stopwords deleted. We used NLTK stopword list for Russian extended with some slang words as кароч "well", типа "like", прост "just", etc.;

Then we applied the following procedures, creating individual datasets for various combinations of them:

1. Leaving or deleting punctuation as tokens;
2. Lemmatization using PyMorphy2 (Korobov, 2015);
3. Stemming using Snowball Stemmer for Russian (Porter, 2001) via NLTK (Bird et al., 2009);

We chose PyMorphy2 over Mystem (Segalovich, 2003) for several reasons, first of them being the time, required to process the data on our OS (Windows). We also prefered pymorphy's lemmatization of unknown words (mostly, names of drugs and medical terms).

3 Method

3.1 Machine Learning

At first, classical Machine Learning models were trained to classify posts on original data without preprocessing. The data was represented as TF-IDF vectors. We decided to proceed with Support Vector Machine with simple linear kernel (LinearSVM), Logistic Regression model (LogReg) and Gradient Boosting Machine (GBM).

3.2 Deep Learning

For this approach, we explored the implementation of Bidirectional Encoder Representations from Transformers (BERT) (Devlin et al., 2018): its extensions for Russian language — RuBert (Kuratov and Arkhipov, 2019) and Conversational RuBERT from DeepPavlov framework (Burtsev et al., 2018). RuBERT had the following characteristics: cased, 12-layer, 768-hidden, 12-heads, 180M parameters, and was fine-tuned with initialization from multilingual BERT on the Russian part of Wikipedia and news data. Conversational RuBERT had the same characteristics and was in turn fine-tuned with RuBERT on OpenSubtitles (Lison and Tiedemann, 2016). Due to the imbalance of classes, as mentioned above — 9% positive to 91% negative, we used following models:

---

[3] `https://github.com/crazyfrogspb/RedditScore`
[4] `https://github.com/aatimofeev/spacy_russian_tokenizer`

Name	Base model	Additions
RuBERT	RuBERT	—
Conv	Conversational RuBERT	—
ConvUnder	Conversational RuBERT	Undersampling
ConvLogReg	Conversational RuBERT	Undersampling + LogReg

Table 1: Model architectures.

For Undersampling approach we split negative class into N equal-sized folds, and combined each split with positive samples, giving that a share of negative samples is almost 0.5. Then we trained N models and stacked their probabilistic predictions for constructing an ensemble architecture. At first, we applied the majority voting method in order to get final answers. Beside that, Logistic Regression model was also used as the ensemble combiner.

4   Experiments

We compared classical ML models with basic BERT models on the original data without any preprocessing. The results reached by classical ML models, which can be found in Table 2, were not competitive, thus we didn't proceed with this approach.

Model	Validation F-score
LinearSVM	0.28
LogReg	0.07
GBM	0.29

Table 2: Classical ML models' results on original data

BERT models showed an increase in F-score, with Conversational RuBERT being slightly ahead of RuBERT. After deciding to stick with the Conversational RuBERT model, we made cross validation in search of the optimal parameters. Further experiments were conducted with Conversational RuBERT model with batch size equal to 32, dropout probability for non-Bert layers 0.4 and learning rate 10^{-5}. All results of testing can be seen in Table 3.

Considering using variants of dataset with additional texts, results were not promising, as there were no visible improvements and the F-score even decreased by 0.05. We believe that the reason for such behaviour lies in the dissimilarities of two types of texts being presumably more significant than we described in Subection 2.1. Implementation of the ensemble architecture proved to be successful and brought up an increase in F-score by 0.05 compared to non-ensemble models when using the ConvLogReg model. It should be noted here that we used a relatively small number of training epochs for models with undersampling, due to the high risk of overfitting.

name	n. models	batch size	n. epochs	last epoch	best epoch
RuBERT	1	64	7	0.36	0.37
Conv	1	64	7	0.45	0.47
ConvUnder	5	32	3	0.45	0.49
			4	0.43	
			7	0.45	
		64	4	0.42	0.46
ConvLogReg	5	64	4	0.49	0.49
	6	32	2	**0.51**	**0.51**
			4	0.48	

Table 3: F-score on validation set for different model configurations.

Further experiments were conducted in order to evaluate which pipeline of data preprocessing suits this task the most. Judging by the results shown in Table 4, simple preprocessing without lemmatization, stemming or punctuation deletion works the best, while any other type of preprocessing leads to a decrease in F-score. We discuss the reasons for that in Subsection 5.1.

prep	del punct	lem	stem	last epoch		best epoch	
				RuBERT	Conv	RuBERT	Conv
no				0.40	0.42	0.41	0.45
yes	no	no	no	0.36	**0.45**	0.37	**0.47**
		yes		0.26	0.34	0.26	0.40
			yes	0.29	0.34	0.29	0.39
	yes	no	no	0.37	0.36	0.38	0.43
		yes		0.33	0.35	0.33	0.38
			yes	0.30	0.37	0.32	0.38

Table 4: F-scores of simple models trained on main data with different preprocessing applied both to training and validation sets.

Given the results above, we settled on using ConvLogReg with 6 splits and small number of epochs. We also submitted one Conv model for comparison. Scores for the final models can be found in Table 5.

model		scores			
		Validation			Test
name	n epochs	precision	recall	F-score	
Conv	6	0.39	0.52	0.45	0.48
ConvLogReg	2	**0.51**	0.45	**0.51**	**0.51**
	4	0.45	**0.58**	0.48	0.50

Table 5: Metrics for three final submissions on validation and test sets.

5 Conclusion

In this work, we have explored an application of Bidirectional Encoder Representations from Transformers (BERT) to the task of text classification in Russian. We have empirically evaluated different versions of tuned RuBERT and preprocessing pipelines against F-score for the "positive" class and experiments have shown that logistic regression trained on the result of a six Conversational RuBERT models ensemble trained on the undersampled data with light preprocessing and tokenized punctuation outperforms every other model, providing a new baseline for ADR presence classification in Russian with F-score 0.51, precision 0.45 and recall 0.60 on the test data.

5.1 Discussion

Our research showed that stemming, lemmatization and punctuation removal when working with Russian language only decreases the final scores. Russian is different from English in such a way that the word order in Russian is free to some extent and syntactic relations are mostly encoded by morphology and punctuation. Knowing that BERT is capable of capturing hierarchy-sensitive and syntactic dependencies (Goldberg, 2019), it becomes obvious, that when dependencies indicators are blurred or removed the results worsen. In addition, some punctuation has its own semantics. For example, "(" means "sad" and "!!!!" can mean high importance. We don't see any premises to remove such kind of data from the dataset. Another point of discussion is the presence of mistakes in the datasets. In the training data we have found some debatable annotations and some which are erroneous for sure. These mistakes could possibly affect the performance of our model.

5.2 Future work

One potential objective for future improvement is to implement rule-based approach to the task, as mixed systems are known to perform better (Ray and Chakrabarti, 2019). We have already made some advancements on this path, but there is still a lot of research to perform. Also, we hope to continue enhancing the results by extending the dataset from Russian social networks and RuDReC corpus.

References

Steven Bird, Ewan Klein, and Edward Loper. 2009. *Natural Language Processing with Python*. 01.

Mikhail Burtsev, Alexander Seliverstov, Rafael Airapetyan, Mikhail Arkhipov, Dilyara Baymurzina, Nikolay Bushkov, Olga Gureenkova, Taras Khakhulin, Yurii Kuratov, Denis Kuznetsov, Alexey Litinsky, Varvara Logacheva, Alexey Lymar, Valentin Malykh, Maxim Petrov, Vadim Polulyakh, Leonid Pugachev, Alexey Sorokin, Maria Vikhreva, and Marat Zaynutdinov. 2018. Deeppavlov: Open-source library for dialogue systems. 07.

Jacob Devlin, Ming-Wei Chang, Kenton Lee, and Kristina Toutanova. 2018. Bert: Pre-training of deep bidirectional transformers for language understanding.

Tilia Ellendorff, Lenz Furrer, Nicola Colic, Noëmi Aepli, and Fabio Rinaldi. 2019. Approaching SMM4H with merged models and multi-task learning. In *Proceedings of the Fourth Social Media Mining for Health Applications (#SMM4H) Workshop & Shared Task*, pages 58–61, Florence, Italy, August. Association for Computational Linguistics.

Yoav Goldberg. 2019. Assessing bert's syntactic abilities. *CoRR*, abs/1901.05287.

Ari Z. Klein, Alimova Ilseyar, Ivan Flores, Arjun Magge, Zulfat Miftahutdinov, Anne-Lyse Minard, Karen O'Connor, Abeed Sarker, Elena Tutubalina, Davy Weissenbacher, and Graciela Gonzalez-Hernandez. 2020. Overview of the fifth social media mining for health applications (#smm4h) shared tasks at coling 2020. *In Proceedings of the Fifth Social Media Mining for Health Applications (#SMM4H) Workshop & Shared Task*.

Mikhail Korobov. 2015. Morphological analyzer and generator for russian and ukrainian languages. In Mikhail Yu. Khachay, Natalia Konstantinova, Alexander Panchenko, Dmitry I. Ignatov, and Valeri G. Labunets, editors, *Analysis of Images, Social Networks and Texts*, volume 542 of *Communications in Computer and Information Science*, pages 320–332. Springer International Publishing.

Yuri Kuratov and Mikhail Arkhipov. 2019. Adaptation of deep bidirectional multilingual transformers for russian language. *CoRR*, abs/1905.07213.

Pierre Lison and Jörg Tiedemann. 2016. OpenSubtitles2016: Extracting large parallel corpora from movie and TV subtitles. In *Proceedings of the Tenth International Conference on Language Resources and Evaluation (LREC'16)*, pages 923–929, Portorož, Slovenia, May. European Language Resources Association (ELRA).

Martin F. Porter. 2001. Snowball: A language for stemming algorithms. Published online, October. Accessed 11.03.2008, 15.00h.

Paramita Ray and Amlan Chakrabarti. 2019. A mixed approach of deep learning method and rule-based method to improve aspect level sentiment analysis. *Applied Computing and Informatics*.

Ilya Segalovich. 2003. A fast morphological algorithm with unknown word guessing induced by a dictionary for a web search engine. In Hamid R. Arabnia and Elena B. Kozerenko, editors, *MLMTA*, pages 273–280. CSREA Press.

Elena Tutubalina, Ilseyar Alimova, Zulfat Miftahutdinov, Andrey Sakhovskiy, Valentin Malykh, and Sergey Nikolenko. 2020. The russian drug reaction corpus and neural models for drug reactions and effectiveness detection in user reviews. *Bioinformatics*, 07. btaa675.

KFU NLP Team at SMM4H 2020 Tasks: Cross-lingual Transfer Learning with Pretrained Language Models for Drug Reactions

Zulfat Miftahutdinov
Kazan Federal University
Kazan, Russia

Andrey Sakhovskiy
Kazan Federal University
Kazan, Russia

Elena Tutubalina
Kazan Federal University
Kazan, Russia

Abstract

This paper describes neural models developed for the Social Media Mining for Health (SMM4H) 2020 shared tasks. Specifically, we participated in two tasks. We investigate the use of a language representation model BERT pretrained on a large-scale corpus of 5 million health-related user reviews in English and Russian. The ensemble of neural networks for extraction and normalization of adverse drug reactions ranked first among 7 teams at the SMM4H 2020 Task 3 and obtained a relaxed F1 of 46%. The BERT-based multilingual model for classification of English and Russian tweets that report adverse reactions ranked second among 16 and 7 teams at two first subtasks of the SMM4H 2019 Task 2 and obtained a relaxed F1 of 58% on English tweets and 51% on Russian tweets.

1 Introduction

Text classification, named entity recognition (NER), and medical concept normalization (MCN) in free-form texts are crucial steps in every text-mining pipeline. Here we focus on discovering adverse drug reaction (ADR) concepts in Twitter messages as part of the Social Media Mining for Health (SMM4H) 2020 shared tasks (Klein et al., 2020).

This work is based on the participation of our team, named *KFU NLP*, in two tasks. Organizers of SMM4H 2020 Task 2 provided participants with train, dev, and test sets of English and Russian tweets annotated at the message level with a binary annotation indicating the presence or absence of ADRs. For Task 3, train, dev, and test sets include tweets with text spans of reported ADRs and their corresponding medical codes from the Medical Dictionary for Regulatory Activities (MedDRA) (Brown et al., 1999).

Neural architectures based on Bidirectional Encoder Representations from Transformers (BERT) (Devlin et al., 2019) have achieved state-of-the-art results in the biomedical domain. For Task 2, we conduct extensive experiments with two SMM4H 2020 train sets separately and a union of these Russian and English sets. For Task 3, we utilize an additional English training sets and dictionaries. We investigate the following versions of BERT:

(1) BioBERT v.1.1 (Lee et al., 2020), pretrained on English texts from PubMed and PMC;

(2) $BERT_{base}$, Multilingual Cased, pretrained on 104 languages, this model was used for the initialization of two models listed below;

(3) EnDR-BERT (Tutubalina et al., 2020), pretrained on the English corpus of 2.6M health-related comments;

(4) EnRuDR-BERT (Tutubalina et al., 2020), pretrained on English and Russian corpora of 5M health-related texts.

51

Proceedings of the 5ᵗʰ Social Media Mining for Health Applications (#SMM4H) Workshop & Shared Task, pages 51–56
Barcelona, Spain (Online), December 12, 2020.

Pretrained weights of domain-specific EnDR-BERT and EnRuDR-BERT models are available at `https://github.com/cimm-kzn/RuDReC`. The source code for our model for Task 2 is available at `https://github.com/Andoree/smm4h_classification`.

The paper is organized as follows. We describe our experiments in the classification of multilingual tweets that report adverse reactions in Section 2. In Section 3, we describe our pipeline for NER and medical concept normalization (MCN) for Task 3. Finally, we discuss future directions in Section 4.

2 Task 2: Classification of Multilingual Tweets

The goal of this task is to detect tweets that report an adverse effect of a medication. We present our results for Russian and English subsets of tweets. The data for Task 2 consists of 3 distinct sets of tweets posted in English, Russian, and French. We present our results on two of them: Russian and English with 20,544 and 6,090 tweets, respectively. The training sets are highly imbalanced as only 1,903 English and 533 Russian tweets are labeled as positive examples. Further, we refer to the union of two train sets as a *bilingual* train. We note that the dev sets are not used for training.

2.1 Models

We utilize three BERT-based models for classification: (i) multilingual $BERT_{base}$[1], (ii) EnDR-BERT[2], and (iii) EnRuDR-BERT[3]. For BERT models, we compared the following training approaches. First, we fine-tuned the mentioned BERT models on training sets. Second, we combined English and Russian subsets of tweets and trained EnRuDR-BERT using their union as a training set. Third, we tried to improve the Russian subtask results using annotated sentences from the RuDReC (Tutubalina et al., 2020) and PsyTAR (Zolnoori et al., 2019) corpora. Annotations include the following types of entities: adverse drug reaction, drug indication, drug effectiveness/ineffectiveness. We pretrained EnRuDR-BERT on the multilabel sentence classification task. Preprocessing of the dataset included the following steps: (i) replacement of all URLs with word "link"; (ii) masking of all @user mentions by @username tag; (iii) mapping of some emojis to the corresponding words (for example, we replaced the pill and syringe emojis with the words pill and syringe); (iv) fix of "&" ampersand representation.

In addition, we use a classification architecture (Kim, 2014) based on convolutional neural networks (CNN) (LeCun et al., 1998) with FastText embeddings (Bojanowski et al., 2017) trained on 1.4 Russian reviews about health & beauty from the RuDReC corpus. The CNN model consisted of 3 convolutional layers with kernel sizes of 3, 4, 5, and ReLU activation. Each convolutional layer consisted of 128 filters and was followed by a max-pooling layer. We used a dense layer with sigmoid activation as the output layer. We used the Keras (Chollet and others, 2015) implementation[4] of the CNN model. For FastText and CNN experiments, we removed punctuation, lowercased, and tokenized all texts using NLTK (Loper and Bird, 2002).

2.2 Experiments

For the classification task, each BERT model was trained for 3 epochs with the learning rate of $3 * 10^{-5}$ using Adam optimizer. We defined the first 10% of the training steps as warm-up steps and set batch size to 64 and maximum sequence size to 128. We utilized Tensorflow (Abadi et al., 2016) implementation of BERT[5] with softmax output activation for fine-tuning and sigmoid output activation for multilabel classification pretraining. We run multilabel classification pretraining for 1 epoch with the learning rate of $2 * 10^{-5}$ and a batch size of 64. We used a classification threshold of 0.5 for all models. We limited the CNN model to 10 epochs with early stopping after 3 epochs with no accuracy improvement on the dev set and set the batch size to 128.

During the evaluation of BERT models, we encountered an instability of results on the dev set. For the final submissions, we used an ensemble of 10 BERT models with a simple voting scheme to solve this

---

[1] `https://github.com/google-research/bert/blob/master/multilingual.md`
[2] `https://huggingface.co/cimm-kzn/endr-bert`
[3] `https://huggingface.co/cimm-kzn/enrudr-bert`
[4] `https://github.com/ShawnyXiao/TextClassification-Keras`
[5] `https://github.com/google-research/bert`

Model	Dev set			Test set (official results)		
	P	**R**	**F1**	**P**	**R**	**F1**
Russian tweets						
Multilingual BERT, train on Russian tweets	0.20	**0.87**	0.32	–	–	–
FastText+CNN, train on Russian tweets	0.28	0.79	0.41	–	–	–
EnRuDR-BERT, train on Russian tweets	0.46	0.37	0.41	–	–	–
EnRuDR-BERT, bilingual train	0.44	0.57	0.50	–	–	–
EnRuDR-BERT, ensemble, bilingual train, pretrained on RuDReC+PsyTAR	**0.55**	0.53	**0.54**	**0.54**	0.48	**0.51**
Average scores provided by organizers	–	–	–	0.36	**0.58**	0.43
English tweets						
EnRuDR-BERT, train on English tweets	**0.65**	0.64	0.64	–	–	–
EnRuDR-BERT, bilingual train	**0.65**	0.65	**0.65**	–	–	–
EnDR-BERT, ensemble	0.60	**0.68**	0.64	**0.63**	0.54	**0.58**
Average scores provided by organizers	–	–	–	0.42	**0.59**	0.46

Table 1: Text classification results on the SMM4H Task 2 dev and test sets.

problem.

Table 1 shows the performance of BERT models and CNN in Task 2 in terms of precision, recall, and F1-score. The following conclusions can be drawn based on the results. First, the FastText + CNN approach outperformed multilingual BERT in terms of F1-score on the Russian subtask. However, EnRuDR-BERT, fine-tuned on Russian tweets, showed a performance comparable to CNN. An explanation for this might be that EnRuDR-BERT and FastText embeddings were pretrained on healthcare domain texts, and domain-specific pretraining considerably increases the quality of fine-tuning. Second, the Russian training set's replacement with the bilingual training set resulted in a significant increase in recall (20%) and F1-score (9%) on the Russian subtask. Third, the addition of the Russian data to the training set does not significantly improve the English subtask's performance. Finally, the combination of multilabel classification pretraining and simple ensembling increased F1-score by 4%.

3 Task 3: Extraction and Normalization of Adverse Reactions

This task's objective is to detect ADR mentions and then map these entities to concepts in a controlled vocabulary.

3.1 Named Entity Recognition

Following the best results in SMM4H 2019 Task 2 & 3 (Miftahutdinov et al., 2019), we utilize a BERT-based model for the named entity recognition task.

We experiment with two initialization checkpoints for the language model: BioBERT and EnDR-BERT. We investigate dictionary-based (gazetteer) features and two strategies of extra training data utilization. Dictionary-based features are calculated for each token in a text as follows: first, all the occurrences of predefined vocabulary entries were found in the text, then the first token of the matched part tagged was with B-tag, the last with I-tag, and all other tokens in the text with O-tag. The dictionary-based features are concatenated with the representation learned by the neural network that captures an entity's extensional semantic information. We adopted the dictionaries from previous work (Miftahutdinov et al., 2017).

As extra training data for the NER task, we used the CSIRO Adverse Drug Event Corpus (CADEC) (Karimi et al., 2015). The first strategy of exploiting the dataset is to train the model on the CADEC data and then fine-tune on the SMM4H train set. The second strategy is to combine these two datasets into one (see '+CADEC' in Table2).

3.2 Normalization

For the normalization task, we applied two models: (i) a classifier (Miftahutdinov et al., 2019; Miftahutdinov and Tutubalina, 2019), (ii) a novel neural model based on similarity distance of BERT vectors of concepts. We also evaluate the combination of the two approaches. Following (Miftahutdinov et al., 2019), we utilize additional data for training. Other corpora are filtered to match a vocabulary of the SMM4H 2020 train set.

Classification over Concepts State-of-the-art studies consider the concept normalization task as a classification problem. We develop a supervised classifier: first, we convert each mention into a vector representation using BERT with mean pooling over layers. Second, we add a softmax layer to convert values to conditional probabilities. The size of the softmax layer is the number of concepts in the terminology.

Metric Learning The second approach is based on metric learning. The intuition here is to map concepts and entities into common embedding space such that entities and their concepts are close to each other. To encode entities, we utilized the BERT model. To encode the concept, first, we extract textual representation from UMLS (Bodenreider, 2004). In particular, UMLS contains concept names of each concept. Each of the concept names is utilized as a textual representation of a concept. Textual representations are then encoded using the BERT model. The entity encoder and the concept name encoder share the weights. Given the embedding u_m of the mention m and u_{c_i} embedding of the i-th concept name c_i, model output is obtained as follows:

$$cid(argmin_{c_i}||u_m, u_{c_i}||),\qquad(1)$$

where cid(*) is a function that maps the concept name to the corresponding concept id. The approach included almost no preprocessing steps except lowercasing entities and concept names.

To encode mentions and concepts, we use BERT model fine-tuned using a triplet loss (Reimers and Gurevych, 2019). Given an user-generated entity mention m, a positive concept name c and a randomly sampled concept name n as negative example, triplet loss tunes the network such that the distance between m and c is smaller than the distance between m and n. Mathematically, we minimize the following loss:

$$max(||u_m, u_c|| - ||u_m, u_n|| + \epsilon, 0)\qquad(2)$$

where u_m, u_c, u_n are the embeddings for m, c, n, $||u_*, u_*||$ a distance metric, ϵ is margin that ensures that u_c is at least ϵ closer to u_m than u_n. As metric, we use Euclidean distance or cosine similarity and we set $\epsilon = 1$ in our experiments.

Combined Method Since the classification approach can't handle out of vocabulary cases we combined two models based on a threshold. For instance, given (i) prediction c_{bs} from from BERT-based similarity method with the distance equals to d and (ii) prediction c_{clf} from the classification approach, the final prediction is set to c_{bs}, if d is less than a threshold, and to c_{clf}, otherwise. Varying the threshold we can customize the model to make predictions based on classification or metric learning approach. With a small threshold, the combined method will rarely output predictions from a metric learning approach and vice versa. It's reasonable to set low threshold values when the training set covers most of the concepts from the test dataset.

3.3 Experiments

For the NER sub-task, each network was trained for 50 epochs with batch size set to 32. We used the Adam algorithm (Kingma and Ba, 2015) as the optimizer with an initial learning rate $5 * 10^{-5}$. We use precision, recall, and F-measure for evaluation. Table 2 shows the performance of the models on the development set. It could be seen that using the fine-tuned language model and additional data gives a significant improvement in F-measure. For both submissions, the model trained using EnDR-BERT, gazetteer features, and extra training data was utilized.

Model	P	R	F1
BioBERT	51.09	54.36	52.68
EnDR-BERT	**58.20**	52.99	55.47
EnDR-BERT, gazetteer features	57.65	53.28	55.38
EnDR-BERT, gazetteer features, CADEC pretrain	57.65	55.53	56.57
EnDR-BERT, gazetteer features, +CADEC	57.65	**57.97**	**57.81**

Table 2: NER performance on the dev set.

Model	Acc@1
EnDR-BERT classifier	42.14
EnDR-BERT classifier, +additional data	44.90
EnDR-BERT, triplet loss	36.91
EnDR-BERT, triplet loss, +additional data	37.19
Combined method, threshold 4.5	45.17
Combined method, threshold 8.2	43.25

Table 3: MCN performance in terms of accuracy@1 on the dev set.

Run name	P	R	F1
ADR Detection Evaluation (Relaxed)			
KFU NLP Team, run 1	79.1	72.3	75.5
KFU NLP Team, run 2	79.1	72.3	75.5
Average scores	60.7	55.7	56.4
End-to-End Evaluation (Relaxed)			
KFU NLP Team, run 1	46.1	42.7	44.3
KFU NLP Team, run 2	48.2	44.6	46.3
Average scores	31.2	29	29.2

Table 4: Performance of our models in SMM4H 2020 Task 3 (official results).

For the normalization task, we trained the classifier based on EnDR-BERT. The classifier is trained for 50 epochs with the Adam optimizer. To train BERT with triplet loss, we generated 109,605 triplets. For each positive concept name, 15 negative examples were generated from the MedDRA vocabulary. The model is trained for 10 epochs. Table 3 shows performance of the normalization models on the development set. The combination of the EnDR-BERT classifier and the BERT-similarity approach with a threshold equal to 4.5 outperformed other models. For the final submission, we selected a combined approach with a threshold equal to 8.2 for run 1 and a combined approach with a threshold equal to 4.5 for run 2.

Table 4 shows a comparison of the model to the official average scores computed using the participants' submissions. Our model has obtained the highest relaxed F1 score of 46.3% and 75.5% on ADR detection and end-to-end tasks, respectively.

4 Conclusion

In this work, we have explored an application of domain-specific BERT models pretrained on health-related user reviews in English and Russian to the task of multilingual text classification and extraction and normalization of adverse drug reactions. Experiments have shown that our BERT models outperform general multilingual BERT and BioBERT on two tasks, achieving the best results in Task 3.

There are several directions for future work arising from the results of our experiments. One potential direction is to investigate the impact of multilingual transfer learning from other corpora on the classification of Russian health-related texts. Another future direction is to verify the efficiency of English data for text classification in languages other than Russian. For Task 2, transfer from English to French remain to be explored.

Acknowledgements

This research was supported by the Russian Science Foundation grant #18-11-00284.

References

Martín Abadi, Ashish Agarwal, Paul Barham, Eugene Brevdo, Zhifeng Chen, Craig Citro, Greg S Corrado, Andy Davis, Jeffrey Dean, Matthieu Devin, et al. 2016. Tensorflow: Large-scale machine learning on heterogeneous distributed systems. *arXiv preprint arXiv:1603.04467*.

Olivier Bodenreider. 2004. The unified medical language system (umls): integrating biomedical terminology. *Nucleic acids research*, 32(suppl_1):D267–D270.

Piotr Bojanowski, Edouard Grave, Armand Joulin, and Tomas Mikolov. 2017. Enriching word vectors with subword information. *Transactions of the Association for Computational Linguistics*, 5:135–146.

Elliot G Brown, Louise Wood, and Sue Wood. 1999. The medical dictionary for regulatory activities (meddra). *Drug safety*, 20(2):109–117.

François Chollet et al. 2015. Keras. https://github.com/fchollet/keras.

Jacob Devlin, Ming-Wei Chang, Kenton Lee, and Kristina Toutanova. 2019. Bert: Pre-training of deep bidirectional transformers for language understanding. In *Proceedings of the 2019 Conference of the North American Chapter of the Association for Computational Linguistics: Human Language Technologies, Volume 1 (Long and Short Papers)*, pages 4171–4186.

Sarvnaz Karimi, Alejandro Metke-Jimenez, Madonna Kemp, and Chen Wang. 2015. Cadec: A corpus of adverse drug event annotations. *Journal of biomedical informatics*, 55:73–81.

Yoon Kim. 2014. Convolutional neural networks for sentence classification. In *Proceedings of the 2014 Conference on Empirical Methods in Natural Language Processing (EMNLP)*, pages 1746–1751.

Diederik Kingma and Jimmy Ba. 2015. Adam: A method for stochastic optimization. *International Conference on Learning Representations 2015*.

Ari Z. Klein, Ilseyar Alimova, Ivan Flores, Arjun Magge, Zulfat Miftahutdinov, Anne-Lyse Minard, Karen O'Connor, Abeed Sarker, Elena Tutubalina, Davy Weissenbacher, and Graciela Gonzalez-Hernandez. 2020. Overview of the fifth social media mining for health applications (# smm4h) shared tasks at coling 2020. In *Proceedings of the Fifth Social Media Mining for Health Applications (# SMM4H) Workshop & Shared Task*.

Yann LeCun, Léon Bottou, Yoshua Bengio, and Patrick Haffner. 1998. Gradient-based learning applied to document recognition. *Proceedings of the IEEE*, 86(11):2278–2324.

Jinhyuk Lee, Wonjin Yoon, Sungdong Kim, Donghyeon Kim, Sunkyu Kim, Chan Ho So, and Jaewoo Kang. 2020. Biobert: a pre-trained biomedical language representation model for biomedical text mining. *Bioinformatics*, 36(4):1234–1240.

Edward Loper and Steven Bird. 2002. Nltk: The natural language toolkit. In *Proceedings of the ACL-02 Workshop on Effective Tools and Methodologies for Teaching Natural Language Processing and Computational Linguistics*, pages 63–70.

Zulfat Miftahutdinov and Elena Tutubalina. 2019. Deep neural models for medical concept normalization in user-generated texts. In *Proceedings of the 57th Annual Meeting of the Association for Computational Linguistics: Student Research Workshop*, pages 393–399.

Z.Sh. Miftahutdinov, E.V. Tutubalina, and A.E. Tropsha. 2017. Identifying disease-related expressions in reviews using conditional random fields. *Komp'juternaja Lingvistika i Intellektual'nye Tehnologii*, 1(16):155–166.

Zulfat Miftahutdinov, Ilseyar Alimova, and Elena Tutubalina. 2019. Kfu nlp team at smm4h 2019 tasks: Want to extract adverse drugs reactions from tweets? bert to the rescue. In *Proceedings of the Fourth Social Media Mining for Health Applications (# SMM4H) Workshop & Shared Task*, pages 52–57.

Nils Reimers and Iryna Gurevych. 2019. Sentence-bert: Sentence embeddings using siamese bert-networks. In *Proceedings of the 2019 Conference on Empirical Methods in Natural Language Processing and the 9th International Joint Conference on Natural Language Processing (EMNLP-IJCNLP)*, pages 3973–3983.

Elena Tutubalina, Ilseyar Alimova, Zulfat Miftahutdinov, Andrey Sakhovskiy, Valentin Malykh, and Sergey Nikolenko. 2020. The russian drug reaction corpus and neural models for drug reactions and effectiveness detection in user reviews. *Bioinformatics*, 07.

Maryam Zolnoori, Kin Wah Fung, Timothy B Patrick, Paul Fontelo, Hadi Kharrazi, Anthony Faiola, Yi Shuan Shirley Wu, Christina E Eldredge, Jake Luo, Mike Conway, et al. 2019. A systematic approach for developing a corpus of patient reported adverse drug events: A case study for ssri and snri medications. *Journal of biomedical informatics*, 90:103091.

SMM4H Shared Task 2020 - A Hybrid Pipeline for Identifying Prescription Drug Abuse from Twitter: Machine Learning, Deep Learning, and Post-Processing

Isabel Metzger[1,5], Emir Y. Haskovic[2], Allison Black[3], Whitley M. Yi[4], Rajat S. Chandra[1],
Mark T. Rutledge[1], William McMahon[1], and Yindalon Aphinyanaphongs[5]

[1]Sumitovant Biopharma, Inc., New York, NY
[2]Lokavant, New York, NY
[3]Roivant Sciences, Inc., New York, NY
[4]Skaggs School of Pharmacy and Pharmaceutical Sciences, University of Colorado, CO
[5]Department of Population Health, NYU Langone Health, New York, NY
{*isabel.metzger, rajat.chandra, mark.rutledge, bill.mcmahon*}*@sumitovant.com,*
emir.haskovic@lokavant.com,
allison.black@roivant.com,
whitley.yi@ucdenver.edu,
yin.a@nyulangone.org

Abstract

This paper presents our approach to multi-class text categorization of tweets mentioning prescription medications as being indicative of potential abuse/misuse (A), consumption/non-abuse (C), mention-only (M), or an unrelated reference (U) using natural language processing techniques. Data augmentation increased our training and validation corpora from 13,172 tweets to 28,094 tweets. We also created word-embeddings on domain-specific social media and medical corpora. Our hybrid pipeline of an attention-based CNN with post-processing was the best performing system in task 4 of SMM4H 2020, with an F1 score of 0.51 for class A.

1 Introduction

Substance abuse is a major public health crisis in the United States (Substance Abuse and Mental Health Services Administration, 2019). Use of natural language processing (NLP) for automatic detection of prescription abuse or misuse in social media posts holds promise as a toxicovigilance strategy for near real-time monitoring of prescription abuse patterns and emerging trends (Hu et al., 2019; Sarker et al., 2020). Research has shown that abuse-indicating social media posts correlate with opioid-related overdose deaths at a geographical level (Sarker et al., 2019). Despite its potential, social media poses significant challenges for detection of self-reported abuse or misuse of prescriptions, including colloquial use of drug names, sparse information, and figurative language. The Social Media Mining for Health Applications (SMM4H) Task 4 challenges participants to classify tweets' mentions of opioids, benzodiazepines, atypical antipsychotics, central nervous system (CNS) stimulants or GABA analogues as potential abuse/misuse (A), consumption (C), mention-only (M) or unrelated (U) (Klein et al., 2020). As patterns of abuse and misuse can vary based on drug class and mechanism of action, our team collaborated with clinical subject matter experts (SMEs) to complete the task.

2 Data and Methods

2.1 Text Pre-processing

Publicly available tweets were pre-processed using the ekphrasis (Baziotis et al., 2017) python package to normalize url, email, percent, money, phone, user, time, and date terms and to annotate hashtags, allcaps, elongated, repeated, emphasis, and censored terms.

Proceedings of the 5th Social Media Mining for Health Applications (#SMM4H) Workshop & Shared Task, pages 57–62
Barcelona, Spain (Online), December 12, 2020.

2.2 Word Representations

We used fastText (Bojanowski et al., 2017) to train a 300 dimensional skip-gram model on a domain-specific corpus of over 6.5 million unique sentences for 500 epochs at a learning rate of 0.025 and a negative sampling loss. Subword character n-gram length were set between 3 and 6 (Bojanowski et al., 2017). Other parameters were set to the suggested values.[1]

Source	Number of Tweets/Sentences
Twitter Stream Archive (archive.org)	3,106,783
Tweets pulled using original drug terms listed in (O'Connor et al., 2020)	85,188
UCI Drug Review datasets (Gräßer et al., 2018)	1,313,299
Consumer Health Question Answer (CHQA) Corpus (Kilicoglu et al., 2018)	2,595
First 1 billion bytes of English Wikipedia [2]	5,244,360

Table 1: Corpora - Text Sources for Unsupervised fastText Word-Embeddings Models

2.3 Drug Lexicon Construction

A gazetteer of drug terms was produced in collaboration with SMEs to aid in feature engineering and data augmentation. We began by selecting the drug terms listed in (O'Connor et al., 2020), and supplemented these with clinically relevant drugs within the same Established Pharmacologic Classes (U.S. Food and Drug Administration, 2018), as well as with other drug classes commonly known to be misused. For each drug term, we compiled a list of its generic and proprietary names along with common misspellings. Additional synonyms and street names for the individual drugs and drug classes (e.g. "benzo" for benzodiazepine) were also identified. Generic and proprietary terms were sourced via SMEs and misspellings were sourced from (Drugs.com, 2020). Select street names were sourced from (Drug Enforcement Administration, 2020). The final gazetteer was composed of 258 terms, representing 54 drugs.

2.4 Data Augmentation

The original training and validation sets were mixed and split into new train and validation sets in a 4:1 ratio. Data augmentation was performed on the new training set with Snorkel[3], which uses a matrix completion-style modeling approach as described in (Ratner et al., 2018). Functions were created following the methods described in (Wei and Zou, 2019), including randomly replacing person names, switching adjectives, and replacing nouns, verbs, and adjectives with synonyms. Additional functions include adding or deleting a random character from tweets and replacing references to family members (e.g. swapping "brother" with "father"). Lastly, functions specific to this use case were also implemented, including replacing a drug name with one from a similar class, replacing verbs that are commonly associated with abuse (e.g switching "inject" with "shoot up"), and replacing verbs that are related to non-abusive consumption (e.g. swapping "prescribe" with "refill"). A random policy was used such that up to four of these transformation functions were applied uniformly at random per tweet if the rule was applicable.

2.5 Support Vector Machine with Feature Engineering

SVMs have been successfully applied to automated classification of text, and are commonly used as baseline for evaluating the effectiveness of new model development (Kowsari et al., 2019). For our baseline we trained an SVM using a linear kernel along with a "one-vs-rest" decision boundary scheme and the following features:

- Vader Sentiment Score[4]
- Binary flag features including presence of commonly abused drug terms, (see 2.3) and presence of emojis commonly associated with drug abuse
- Count-based features such as number of hashtags, username mentions, emojis/emoticons and number of drug terms identified (see 2.3)

---

[1] https://fasttext.cc/docs/en/options.html
[2] https://github.com/facebookresearch/fastText/blob/master/get-wikimedia.sh
[3] https://www.snorkel.org
[4] Vader compound score was used to measure sentiment https://github.com/cjhutto/vaderSentiment

- Co-occurrence of chemical and disease entities, where entity extraction was performed using en_ner_bc5cdr_md[5]
- tf-idf of tokens (unigrams and bigrams)

2.6 Convolutional Neural Network Architectures

Convolutional neural networks (CNNs) have shown success in short text classification, especially when orthology is important, and is computationally efficient (Kim, 2014; Minaee et al., 2020). Pre-training was performed with the following parameters (conv depth of 4, cosine loss function, drop out of 0.4, and batch size of 64) on the raw Task 4 dataset with the word embeddings from section 2.2 to build "token to vector" (tok2vec) weights. The tok2vec layer initializes the embedding matrix. The embedding layer also embeds linguistic features such as shape, suffix and prefix. We initially investigated using a 4-layer CNN with mean-pooling and softmax. In our final and winning system, we replaced the 4-layer CNN with a 2-layer hierarchical CNN and used a parametric attention layer as described in (Yang et al., 2016). The resulting weighted summary vector was then passed through a multi-layer perceptron, where each neuron was stacked with a unigram logistic regression to produce four label predicted probabilities similar to the output layer in the attention-based CNN in (Yin et al., 2015). The final classification model architecture is illustrated in figure 1. Replacing a non-linearity with a logistic regression has been shown to be effective (Zhining et al., 2016; Yin et al., 2015). We used a compounding batch size that was trained for 32 epochs. Our post-processing pipeline (2.7) was then applied to the predictions to produce the final submission.

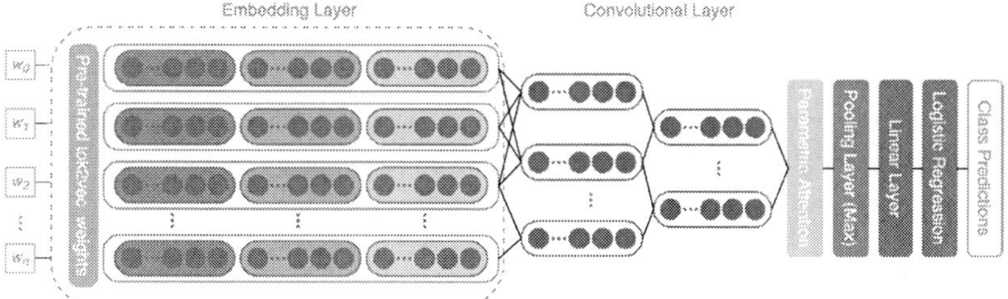

Figure 1: Attention Based CNN with Output Layer

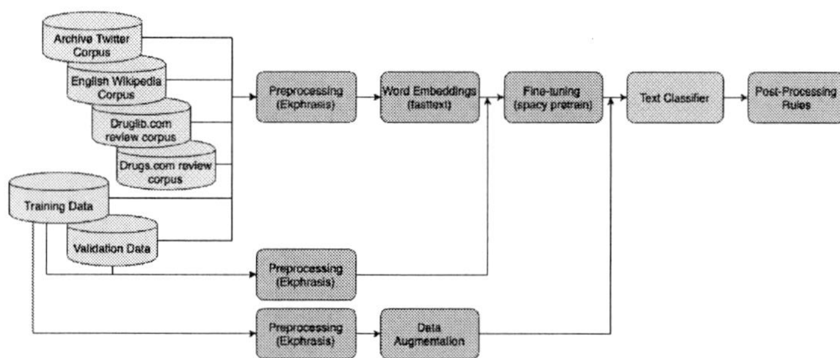

Figure 2: Processing pipeline

2.7 Label Functions and Post-processing

Heuristic post-processing rules were created in conjunction with SMEs and transcribed into Snorkel[3] labelling functions. Rule performance was evaluated on the validation data, and a subset of rules with an empirical accuracy of 0.65 or above were selected and ordered by relative precision. These were then applied to tweets after being fed through the text classifier. For each tweet the rules are applied in succession until a rule "votes" to set the prediction label, at which point no further rules are applied and the tweet's label is updated. If no rules vote then the label provided by the classifier is unchanged.

---

[5]https://allenai.github.io/scispacy/

	j	Polarity	Coverage	Overlaps	Conflicts	Correct	Incorrect	Emp. Acc.
drug_with_slang_usage	0	[0]	0.019645	0.019645	0.019645	87	120	0.420290
drug_with_consumption_usage	1	[0]	0.100978	0.100978	0.100978	210	854	0.197368
no_drugnames_found	2	[3]	0.004176	0.004176	0.004176	6	38	0.136364
...								
drug_had_me	30	[2]	0.011958	0.011958	0.011958	87	39	0.690476
died_of	31	[1]	0.001139	0.001139	0.001139	10	2	0.833333
lil_xan	32	[1, 3]	0.002373	0.002373	0.002373	21	4	0.840000
addicted_3rdperson	33	[0, 1, 2]	0.031603	0.031603	0.031603	212	121	0.636637

(a) Subset of rules and their empirical scores

```
-- Pet Meds labelling rule
--   Identify cases where medications are mentioned
--   as being used for pet-related consumption.
--   Look for pet terms that occur before a drug
--   term in the tweet, e.g. "I give my puppy some
--   xanax before going to the vet."
rule Pet_Meds
  set vote to 'PASS';
  For i := 2 to length(tweet_tokens) do
   If tweet_tokens[i] in drug_gazetteer then
     For j := 1 to i-1 do
       If tweet_tokens[j] in pet_terms then
         set vote to 'MENTION';
  end;
```

(b) Example rule pseudo-code

Figure 3: Labelling functions evaluated for post-processing

3 Discussion

3.1 Error Analysis

Our hybrid pipeline commonly labeled abuse (A) tweets as mentions (M) and to a lesser degree as consumption (C). Manual review showed the model struggled to detect abuse when the subject of the abuse was of a relative or close friend, which was included as a criteria for the definition of abuse. This may be due to the low prevalence of family member- or relative-related abuse in the training dataset. Among abuse tweets mislabeled as mention from a subset of the Task 4 validation set, the most common medication class was benzodiazepines (33/78), followed by opioids (20/78) and then CNS stimulants (19/78). Of the benzodiazepines, the most common drug among the mislabeled abuse tweets was Xanax, which could be due to its popularity in everyday vernacular, making it more challenging for an NLP model to delineate subtle differences in tweets' semantic meaning. For example, "xanax" has been used as an adjective to describe individuals or a group or used as a figure of speech to express the need to calm down (see examples below).

- If one more entitled Xanax mom yells at me to honor a coupon that expired 3 years ago I'm jumping off a cliff
- honestly i feel like walkin around with xanax and just throwing it at people cause ya just needa chilll

	precision	recall	f1-score	support
ABUSE	0.732	0.634	0.679	448
CONSUMPTION	0.781	0.858	0.817	730
MENTION	0.882	0.876	0.879	1353
UNRELATED	0.931	0.913	0.922	104

Table 2: Detailed Performance of final CNN architecture on validation split + augmented validation data

4 Results and Conclusion

In this paper, we evaluate two systems as shown in Table 3.

model system	F1 Score	Precision	Recall
SVM + Feature engineering	0.49	0.5276	0.4553
CNN with Attention and stacked Linear + post-processing	0.51	0.5306	0.4831

Table 3: System performance on official test data

Opportunities to improve upon current work include development of additional post-processing rules to improve labeling for edge cases and further refinement of the heuristics with the guidance of SMEs. Directions for future work in the field of NLP and prescription drug abuse include exploration of alternative methods to disambiguate ground truth labels. Due to the sparsity of contextual information available to confirm the suspicion of abuse or misuse in an individual tweet, a future strategy is to consider the creation of confidence levels for ground truth abuse/misuse labels to differentiate between high confidence of abuse versus subjective hints of abuse.

References

Christos Baziotis, Nikos Pelekis, and Christos Doulkeridis. 2017. DataStories at SemEval-2017 task 4: Deep LSTM with attention for message-level and topic-based sentiment analysis. In *Proceedings of the 11th International Workshop on Semantic Evaluation (SemEval-2017)*, pages 747–754, Vancouver, Canada, August. Association for Computational Linguistics.

Piotr Bojanowski, Edouard Grave, Armand Joulin, and Tomas Mikolov. 2017. Enriching word vectors with subword information. *Transactions of the Association for Computational Linguistics*, 5:135–146.

Drug Enforcement Administration. 2020. *Drugs of Abuse, A DEA Resource Guide*. Drug Enforcement Administration.

Drugs.com. 2020. Phonetic and wildcard search at drugs.com. `https://www.drugs.com/search-wildcard-phonetic.html`.

Felix Gräßer, Surya Kallumadi, Hagen Malberg, and Sebastian Zaunseder. 2018. Aspect-based sentiment analysis of drug reviews applying cross-domain and cross-data learning. In *Proceedings of the 2018 International Conference on Digital Health*, DH '18, page 121–125, New York, NY, USA. Association for Computing Machinery.

Han Hu, Nhathai Phan, Soon A Chun, James Geller, Huy Vo, Xinyue Ye, Ruoming Jin, Kele Ding, Deric Kenne, and Dejing Dou. 2019. An insight analysis and detection of drug-abuse risk behavior on twitter with self-taught deep learning. *Computational Social Networks*, 6(1):10, November.

Halil Kilicoglu, Asma Ben Abacha, Yassine Mrabet, Sonya E. Shooshan, Laritza Rodriguez, Kate Masterton, and Dina Demner-Fushman. 2018. Semantic annotation of consumer health questions. *BMC Bioinformatics*, 19(1).

Yoon Kim. 2014. Convolutional neural networks for sentence classification.

Ari Z. Klein, Ilseyar Alimova, Ivan Flores, Arjun Magge, Zulfat Miftahutdinov, Anne-Lyse Minard, Karen O'Connor, Abeed Sarker, Elena Tutubalina, Davy Weissenbacher, and Graciela Gonzalez-Hernandez. 2020. Overview of the fifth Social Media Mining for Health Applications (#SMM4H) Shared Tasks at COLING 2020. In *Proceedings of the Fifth Social Media Mining for Health Applications (#SMM4H) Workshop Shared Task*.

Kowsari, Jafari Meimandi, Heidarysafa, Mendu, Barnes, and Brown. 2019. Text classification algorithms: A survey. *Information*, 10(4):150, Apr.

Shervin Minaee, Nal Kalchbrenner, Erik Cambria, Narjes Nikzad, Meysam Chenaghlu, and Jianfeng Gao. 2020. Deep learning based text classification: A comprehensive review.

Karen O'Connor, Abeed Sarker, Jeanmarie Perrone, and Graciela Gonzalez Hernandez. 2020. Promoting reproducible research for characterizing nonmedical use of medications through data annotation: Description of a twitter corpus and guidelines. *J Med Internet Res*, 22(2):e15861, Feb.

Alexander Ratner, Braden Hancock, Jared Dunnmon, Frederic Sala, Shreyash Pandey, and Christopher Ré. 2018. Training complex models with multi-task weak supervision.

Abeed Sarker, Graciela Gonzalez-Hernandez, Yucheng Ruan, and Jeanmarie Perrone. 2019. Machine learning and natural language processing for geolocation-centric monitoring and characterization of opioid-related social media chatter. *JAMA Network Open*, 2(11).

Abeed Sarker, Annika DeRoos, and Jeanmarie Perrone. 2020. Mining social media for prescription medication abuse monitoring: a review and proposal for a data-centric framework. *J. Am. Med. Inform. Assoc.*, 27(2):315–329, February.

Substance Abuse and Mental Health Services Administration. 2019. Key substance use and mental health indicators in the united states: Results from the 2018 national survey on drug use and health (HHS publication no. pep19-5068, nsduh series h-54). Technical report, Center for Behavioral Health Statistics and Quality, Substance Abuse and Mental Health Services Administration, Rockville, MD.

U.S. Food and Drug Administration. 2018. Pharmacologic class. `https://www.fda.gov/industry/structured-product-labeling-resources/pharmacologic-class`.

Jason W. Wei and Kai Zou. 2019. EDA: easy data augmentation techniques for boosting performance on text classification tasks. *CoRR*, abs/1901.11196.

Zichao Yang, Diyi Yang, Chris Dyer, X. He, Alex Smola, and E. Hovy. 2016. Hierarchical attention networks for document classification. In *HLT-NAACL*.

Wenpeng Yin, Hinrich Schütze, Bing Xiang, and Bowen Zhou. 2015. ABCNN: attention-based convolutional neural network for modeling sentence pairs. *CoRR*, abs/1512.05193.

Lang Zhining, Gu Xiaozhuo, Zhou Quan, and Xu Taizhong. 2016. Combining statistics-based and cnn-based information for sentence classification. *2016 IEEE 28th International Conference on Tools with Artificial Intelligence (ICTAI)*, pages 1012–1018.

Automatic Detecting for Health-related Twitter Data with BioBERT

Yang Bai, Xiaobing Zhou*
School of Information Science and Engineering
Yunnan University, Yunnan, P.R. China
*Corresponding author:zhouxb@ynu.edu.com

Abstract

Social media used for health applications usually contains a large amount of data posted by users, which brings various challenges to NLP, such as spoken language, spelling errors, novel/creative phrases, etc. In this paper, we describe our system submitted to SMM4H 2020: Social Media Mining for Health Applications Shared Task which consists of five sub-tasks(Ari Z. Klein and Gonzalez-Hernandez., 2020). We participate in subtask 1, subtask 2-English, and subtask 5. Our final submitted approach is an ensemble of various fine-tuned transformer-based models. We illustrate that these approaches perform well in imbalanced datasets (For example, the class ratio is 1:10 in subtask 2), but our model performance is not good in extremely imbalanced datasets (For example, the class ratio is 1:400 in subtask 1). Finally, in subtask 1, our result is lower than the average score, in subtask 2-English, our result is higher than the average score, and in subtask 5, our result achieves the highest score.

1 Introduction

According to the United States Centers for Disease Control and Prevention (CDC), drugs administered for alleviating common sufferings are the fourth biggest cause of death and birth defects are the leading cause of infant mortality. These are the most important medical problems for human society.(Giacomini et al., 2007)

Twitter, a popular micro-blogging service, has received much attention recently. It is an online network used by millions of people around the world to stay connected to their friends, family members, and co-workers through their computers and mobile telephones(Barbosa and Feng, 2010). On average, one in a thousand messages from public Twitter data is health-related. These health-related Twitter posts can help us analyze various human health-related phenomena. For example, there are limited methods for studying birth defects in infants, and this knowledge has been challenging(Klein et al., 2018)(Klein et al., 2019). This situation provides a challenging opportunity due to the increasing number of related tweets on Twitter.

With this motivation, five shared tasks are conducted as part of the Social Media Mining for Health Applications (SMM4H) Workshop 2020 hosted by the University of Pennsylvania Health Language Processing (HLP) Lab. Our team participated in subtasks 1, 2-English, and 5 of the workshop. Some samples from the training set are given in Table 1, and these tasks are:

- Sub-task 1: Automatic classification of tweets that mention medications

  We take subtask-1 as a binary classification task. The model is required to determine whether a medication or dietary supplement is mentioned in a tweet, and predict a label L, where $L \in$ {mention a medication or dietary supplement - 1, no mention - 0}.

- Sub-task 2-English: Automatic classification of English tweets that report adverse effects

Proceedings of the 5th Social Media Mining for Health Applications (#SMM4H) Workshop & Shared Task, pages 63–69
Barcelona, Spain (Online), December 12, 2020.

We take subtask-2-English as a binary classification task. The model is required to determine whether an adverse effect of medication is reported in a tweet, and predict a label L, where $L \in \{$ report adverse effects of medication - 1, no report - 0$\}$.

- Sub-task 5: Automatic classification of tweets reporting a birth defect pregnancy outcome

  We take subtask-5 as a ternary classification task. The model is required to determine whether the user has a child and indicate that the child has the birth defect mentioned in a tweet, and predict a label L, where $L \in \{$ refer to the users child and indicate that he/she has the birth defect mentioned in the tweet - 1, ambiguous about whether someone is the users child and/or has the birth defect mentioned in the tweet - 2, merely mention birth defects - 3$\}$.

Sub-task	Sentence	Gold Label
Sub-task 1	@username you can try vitamins for Olivia! Hudson has taken vitamins since the NICU & it helps him gain weight	1
	Can someone get me some chippy chips please	0
Sub-task 2 english	today i'm an emotional mess. that's what happens when i think i could wean myself off of cymbalta	1
	@username. prozac makes us all better people. #prozacnation	0
Sub-task 5	@uesername, my baby was born with microcephaly in the US due to #cytomegalovirus not #zika. Why no warning about that? #CDCchat"	1
	"Username @Usrename in a way, mine are advantaged by the older having Down's syndrome. I can recommend it!"	2
	Girl born without an ear can finally hear properly for the first time after surgeons rebuild it	3

Table 1: Samples from the training set of three subtasks

We try a variety of approaches on these tasks, including classical machine learning methods, CNN models, RNN models, and transformer-based models. We find that the transformer-based neural models consistently outperform other methods. The framework of our implementation is shown in Figure 1. Firstly, we combine the official training set and the validation set to get the new data set, which is split into the new training set and the validation set by using the stratified 5-fold cross-validation[1]. Secondly, the test set is predicted by fine tuning the model. Thirdly, we create pseudo-label to combine training set and input these data into model training and prediction. Finally, we get the final result by hard voting.

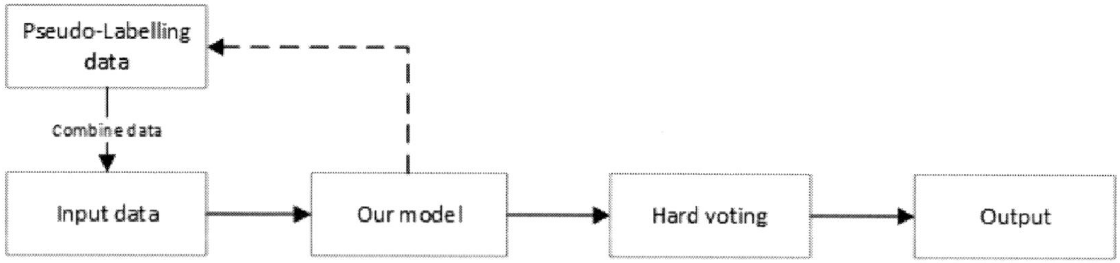

Figure 1: Our framework

2 Related Work

Normally, data in the health field is difficult to obtain. Fortunately, the existence of social media such as Twitter has alleviated this situation. However, it has caused problems that the data is difficult to handle.

---

[1]https://scikit-learn.org/stable/modules/generated/sklearn.model_selection.StratifiedKFold.html

Because diseases and other health problems belong to a small number of people. This phenomenon has caused data imbalance in social media. These problems bring great challenges.

Much previous work has focused on tracking and monitoring diseases on social media. (Yin et al., 2015) has developed a scalable system by training classifiers on a dataset of 34 health topics. It created a general health classifier using standard SVM. Recently, the use of dense word vectors or embeddings is becoming popular, while previous models (e.g. Word2Vec (Pennington et al., 2014), GloVe (Pennington et al., 2014), FastText (Joulin et al., 2016)) focus on learning context-independent word representations, recent works have focused on learning context-dependent word representations. For instance, BERT (Devlin et al., 2018) is a contextualized word representation model, which is based on a masked language model and pre-trained by using bidirectional transformers(Weissenbacher et al., 2018; Weissenbacher et al., 2019b). BioBERT is a domain-specific language representation model pre-trained on a large biomedical corpus(Lee et al., 2020). In this paper, we use the BioBERT in our experiments and use it for the different text classification tasks of Social Media Mining for Health Workshop. In the following, we describe our dataset and the methods used for different tasks.

3 Methodology

3.1 Dataset

The data sets of these tasks are all from social media. These data sets contains rare health-related events such as pregnant women groups, drug effects, birth defects, etc(Sarker et al., 2017; Weissenbacher et al., 2019a; O'Connor et al., 2020).

- For shared task 1. In the training set, 55,419 tweets are included, with 146 tweets mentioning medications (1) and 55,273 tweets not mentioning (0). In the validation set, 13853 tweets are included, 35 labeled 1, and 13818 tweets labeled 0. This is an extremely imbalanced data set, so the evaluation indicator for this task is F1-score for the positive class (i.e., tweets that mention medications).

- For shared task 2. In the training set, 20,544 tweets are included, 1903 tweets that report adverse effects of medications (1), and 18461 tweets that do not report (0). In the validation set, 5134 tweets are included, 474 tweets that report adverse effects of medications (1), and 4660 tweets that do not report (0). This is an imbalanced data set, so the evaluation indicator for this task is F1-score for the positive class (i.e., tweets that report adverse effects of medications).

- For shared task 5. In the training set, there are 14717 tweets, 773 tweets for 'defect' (1), 834 tweets for 'possible defect' class (2), and 13110 tweets for 'non-defect' (2). In the validation set, there are 3680 tweets, 193 tweets for 'defect' (1), 207 tweets for 'possible defect' (2), and 3280 tweets for 'non-defect' (3). This is an imbalanced data set, so the evaluation indicator for this task is micro-averaged F1-score for the "defect" and "possible defect" classes.

3.2 Models

There are some limitations in applying NLP directly to biomedical text mining. With the recent word representation model (such as word2vec), Elmo and BERT are trained and tested on data sets containing common domain text. However, these models do not perform well on biomedical text data sets. So, we choose the BioBERT[2] as our model for these tasks we participated in. For classification tasks, the output of BioBERT(pooler output) is obtained by its last layer hidden state of the first token of the sequence (CLS token) further processed by a linear layer and a tank activation function. But the pooler output is usually not a good summary of the semantic content of the input. So we try the following model architecture to relieve this problem. Our model architecture is shown in Figure 2.

For Figure 2 (a), we can regard the model as two parts. The first part is to get the output of BioBERT(P_O). The second part is the BiGRU module with input P_O. For Figure 2(b), we concatenate P_O and H_0 of the last two hidden layers into the classifier after obtaining P_O. For Figure

---

[2]https://github.com/vthost/biobert-pretrained-pytorch/releases/tag/v1.1-pubmed

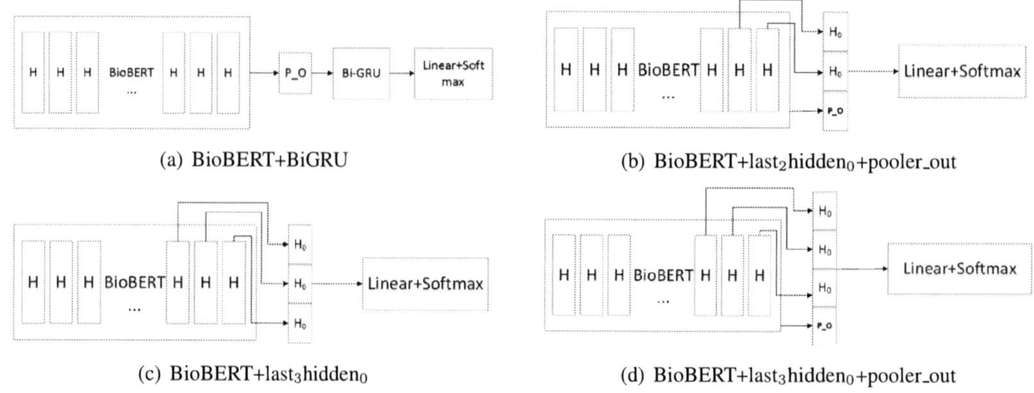

(a) BioBERT+BiGRU   (b) BioBERT+last$_2$hidden$_0$+pooler_out

(c) BioBERT+last$_3$hidden$_0$ (d) BioBERT+last$_3$hidden$_0$+pooler_out

Figure 2: The models we used (H is the hidden layer of BioBERT, P_O is the pooler output, H_0 is hidden-state of the first token of the sequence(CLS token) at the output of the hidden layer of the model.)

2(c), we concatenate H_0 of the last two hidden layers into the classifier. For Figure 2(d), we concatenate P_O and H_0 of the last three hidden layers into the classifier after obtaining P_O.

3.3 Pseudo Label

In order to make our model more robust, we use pseudo labels. First, we input the training set and the validation set into the $model A$ for training, and predict a $result A$ in the test set. Secondly, we add 10% of $result A$ to the training set to obtain a training set with pseudo-label. Again, we input the training with pseudo-label and validation set into the $model A$ for training, and predict a final result in the test set. The process of pseudo label is shown in Figure 3.

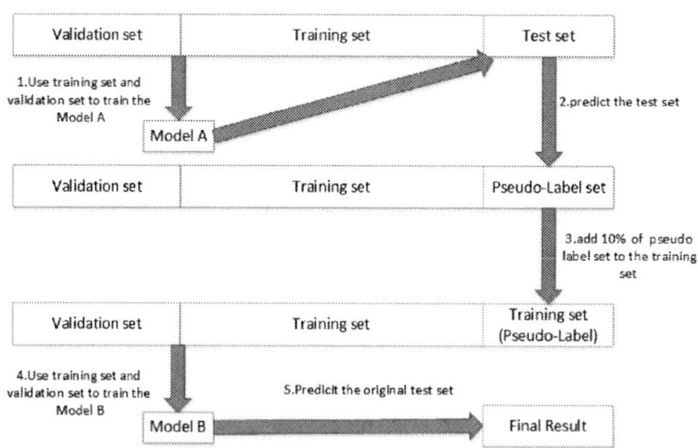

Figure 3: The process of pseudo label

4 Experiments Result

4.1 Experiments Setup

BioBERT (Bidirectional Encoder Representations from Transformers for Biomedical Text Mining) is a domain-specific language representation model pre-trained on large-scale biomedical corpora. It greatly outperforms BERT and previous advanced models in many biomedical text mining tasks. For all tasks, we use the BioBERT-Base-v1.1-PubMed model and RAdam(Liu et al., 2019) as BioBERT optimizer. In order to increase the difference in model fusion, we preprocess the data input by some models. Table 2 shows the hyper-parameters used by our different models. Preprocessing data is simple data cleaning of

	Models	Hyperparameters
1	5-fold data with 42 random seeds BioBERT+BiGRU	output_hidden_states=False dropout=0.1 learning rate=3e-5 epoch=3 per_gpu_train_batch_size=4 gradient_accumulation_steps=4
2	5-fold data with 42 random seeds BioBERT+$\text{last}_2\text{hidden}_0$+pooler_out	output_hidden_states=True dropout=0.1 learning rate=3e-5 epoch=3 per_gpu_train_batch_size=4 gradient_accumulation_steps=4
3	5-fold data with42 random seeds BioBERT+$\text{last}_3\text{hidden}_0$	
4	5-fold data with 24 random seeds BioBERT+BiGRU	
5	5-fold data with 24 random seeds BioBERT+$\text{last}_2\text{hidden}_0$+pooler_out	
6	5-fold data with 42 random seeds Preprocess data BioBERT+$\text{last}_2\text{hidden}_0$+pooler_out	
7	5-fold data with 42 random seeds Pseudo Label Preprocess data BioBERT+BiGRU	
8	5-fold data with 42 random seeds Pseudo Label Preprocess data BioBERT+$\text{last}_2\text{hidden}_0$+pooler_out	
9	5-fold data with 42 random seeds BioBERT+$\text{last}_3\text{hidden}_0$+pooler_out	output_hidden_states=True dropout=0.1 learning rate=2e-5 epoch=3 per_gpu_train_batch_size=4 gradient_accumulation_steps=4

Table 2: Hyperparameters of these models in our experiments for three subtasks.

the data set such as: removing URL, standardizing case, and standardizing @ username. The purpose of preprocessing data is to increase the difference of results during model ensemble.

From table 2, we can see the Hyperparameters of these models. Firstly, we combine the official training set and the verification set to get the new data set, which is split into the new training set and the verification set by using the stratified 5-fold cross-validation. Secondly, we input data into one to six models for training with the training set and predict six results with the test set. We combine these six results($result1 - 6$) to get $result - A$ by hard voting. Thirdly, in order to make pseudo label, we take 10% of the data set from $result - A$ which is combined into the training set. We input the data set with pseudo-label into the seven-eight model to train and predict, we get these result ($result7 - 8$). Fourthly, we input the data into the ninth model training and prediction, we get the result($result9$). Finally, we combine $result1 - 9$ by hard voting to get the final result.

4.2 Results

Table 3 presents the performance scores for three subtasks on the test data, our results are based on hard voting of the 9 models in Table 2. For all tasks, we only use the official training set and validation set and do not use any external data. As can be seen from the table, in extremely imbalanced data, our method

		Precision	Recall	F-1 Score
Subtask-1	Our score	0.8649	0.4156	0.5614
	Mean score	0.7007	0.7039	0.6646
Subtask-2 english	Our score	0.49	0.60	0.54
	Mean score	0.42	0.59	0.46
Subtask-5	Our score	0.65	0.73	0.69
	Mean score	0.62	0.68	0.65

Table 3: The results of our method on the test set for three subtasks

has no advantage. But in imbalanced data, our method has achieved good results.

5   Conclusion

In this work, our method is an ensemble of various fine-tuned transformer-based models on these tasks. We obtain decent results for these tasks organized as a shared task in Social Media Mining for Health Workshop - 2020. Our biggest regret in this work is that in extremely imbalanced data sets, we have not done too much processing on the data sets and can't achieve promising results. In the future, our work will focus on solving the problem of extremely imbalanced data. The code is available online.[3]

References

Ivan Flores Arjun Magge Zulfat Miftahutdinov Anne-Lyse Minard Karen O'Connor Abeed Sarker Elena Tutubalina Davy Weissenbacher Ari Z. Klein, Ilseyar Alimova and Graciela Gonzalez-Hernandez. 2020. Overview of the fifth social media mining for health applications (#SMM4H) shared tasks at COLING 2020. in proceedings of the fifth social media mining for health applications (#SMM4H) Workshop & Shared Task.

Luciano Barbosa and Junlan Feng. 2010. Robust sentiment detection on Twitter from biased and noisy data. In *Coling 2010: Posters*, pages 36–44.

Jacob Devlin, Ming-Wei Chang, Kenton Lee, and Kristina Toutanova. 2018. BERT: Pre-training of deep bidirectional transformers for language understanding. *arXiv preprint arXiv:1810.04805*.

Kathleen M Giacomini, Ronald M Krauss, Dan M Roden, Michel Eichelbaum, Michael R Hayden, and Yusuke Nakamura. 2007. When good drugs go bad. *Nature*, 446(7139):975–977.

Armand Joulin, Edouard Grave, Piotr Bojanowski, and Tomas Mikolov. 2016. Bag of tricks for efficient text classification. *arXiv preprint arXiv:1607.01759*.

Ari Z Klein, Abeed Sarker, Haitao Cai, Davy Weissenbacher, and Graciela Gonzalez-Hernandez. 2018. Social media mining for birth defects research: a rule-based, bootstrapping approach to collecting data for rare health-related events on Twitter. *Journal of biomedical informatics*, 87:68–78.

Ari Z Klein, Abeed Sarker, Davy Weissenbacher, and Graciela Gonzalez-Hernandez. 2019. Towards scaling Twitter for digital epidemiology of birth defects. *NPJ digital medicine*, 2(1):1–9.

Jinhyuk Lee, Wonjin Yoon, Sungdong Kim, Donghyeon Kim, Sunkyu Kim, Chan Ho So, and Jaewoo Kang. 2020. Biobert: a pre-trained biomedical language representation model for biomedical text mining. *Bioinformatics*, 36(4):1234–1240.

Liyuan Liu, Haoming Jiang, Pengcheng He, Weizhu Chen, Xiaodong Liu, Jianfeng Gao, and Jiawei Han. 2019. On the variance of the adaptive learning rate and beyond. *arXiv preprint arXiv:1908.03265*.

Karen O'Connor, Abeed Sarker, Jeanmarie Perrone, and Graciela Gonzalez Hernandez. 2020. Promoting reproducible research for characterizing nonmedical use of medications through data annotation: Description of a Twitter corpus and guidelines. *Journal of medical Internet research*, 22(2):e15861.

Jeffrey Pennington, Richard Socher, and Christopher D Manning. 2014. Glove: Global vectors for word representation. In *Proceedings of the 2014 conference on empirical methods in natural language processing (EMNLP)*, pages 1532–1543.

---

[3]https://github.com/byew/SMM4H-2020

Abeed Sarker, Pramod Chandrashekar, Arjun Magge, Haitao Cai, Ari Klein, and Graciela Gonzalez. 2017. Discovering cohorts of pregnant women from social media for safety surveillance and analysis. *Journal of medical Internet research*, 19(10):e361.

Davy Weissenbacher, Abeed Sarker, Michael Paul, and Graciela Gonzalez. 2018. Overview of the third social media mining for health (SMM4H) shared tasks at EMNLP 2018. In *Proceedings of the 2018 EMNLP Workshop SMM4H: The 3rd Social Media Mining for Health Applications Workshop & Shared Task*, pages 13–16.

Davy Weissenbacher, Abeed Sarker, Ari Klein, Karen OConnor, Arjun Magge, and Graciela Gonzalez-Hernandez. 2019a. Deep neural networks ensemble for detecting medication mentions in tweets. *Journal of the American Medical Informatics Association*, 26(12):1618–1626.

Davy Weissenbacher, Abeed Sarker, Arjun Magge, Ashlynn Daughton, Karen OConnor, Michael Paul, and Graciela Gonzalez. 2019b. Overview of the fourth social media mining for health (SMM4H) shared tasks at ACL 2019. In *Proceedings of the Fourth Social Media Mining for Health Applications (# SMM4H) Workshop & Shared Task*, pages 21–30.

Zhijun Yin, Daniel Fabbri, S Trent Rosenbloom, and Bradley Malin. 2015. A scalable framework to detect personal health mentions on Twitter. *Journal of medical Internet research*, 17(6):e138.

Exploring Online Depression Forums via Text Mining: A Comparison of Reddit and a Curated Online Forum

Luis Moßburger
Media Informatics Group
University of Regensburg
Regensburg, Germany
luis1.mossburger
@stud.uni-regensburg.de

Felix Wende
Media Informatics Group
University of Regensburg
Regensburg, Germany
felix.wende
@stud.uni-regensburg.de

Kay Brinkmann
Media Informatics Group
University of Regensburg
Regensburg, Germany
kay.brinkmann
@stud.uni-regensburg.de

Thomas Schmidt
Media Informatics Group
University of Regensburg
Regensburg, Germany
thomas.schmidt@ur.de

Abstract

We present a study employing various techniques of text mining to explore and compare two different online forums focusing on depression: (1) the subreddit r/depression (over 60 million tokens), a large, open social media platform and (2) Beyond Blue (almost 5 million tokens), a professionally curated and moderated depression forum from Australia. We are interested in how the language and the content on these platforms differ from each other. We scrape both forums for a specific period. Next to general methods of computational text analysis, we focus on sentiment analysis, topic modeling and the distribution of word categories to analyze these forums. Our results indicate that Beyond Blue is generally more positive and that the users are more supportive to each other. Topic modeling shows that Beyond Blue's users talk more about adult topics like finance and work while topics shaped by school or college terms are more prevalent on r/depression. Based on our findings we hypothesize that the professional curation and moderation of a depression forum is beneficial for the discussion in it.

1 Introduction

In recent years, online forums and communities have become an important outlet for growing numbers of people who struggle with depression. For example, the subreddit *r/depression* on Reddit had a growth from around 100,000 members in 2015 to over 570,000 in 2020[1]. While a few decades ago, affected would have had to find support or self-help groups in their local area, they are now able to talk about depression and share their stories with thousands of like minded individuals from the comfort of their own home. Besides r/depression, there are traditional forums focusing on mental health that offer people a place to talk about their condition. Some of these are professionally curated as well, providing support hotlines, ties to experts in psychiatry and further articles on important topics.

In the following paper, we want to investigate how language and content of a platform like r/depression differs from a traditional and professionally accompanied forum and how we can explore these questions with the support of text mining methods. Specifically, we compare content and language through various text mining methods, using word frequencies, topic modeling, word categories as well as sentiment analysis. We regard the project at the moment as descriptive and explorative, therefore we do not want to validate concrete hypothesis but rather explore methods and data to formulate potential hypotheses for future work. While there is research applying computational text analysis on depression forums, we are not aware of similar research focusing on the comparison of a curated and a non-curated forum.

[1] https://subredditstats.com/r/depression

Proceedings of the 5th Social Media Mining for Health Applications (#SMM4H) Workshop & Shared Task, pages 70–81
Barcelona, Spain (Online), December 12, 2020.

2 Related Work

First, we shortly describe the main text mining methods we apply in this study: Sentiment analysis is the computational method to analyze the sentiment expressed towards entities, mostly in written text (Liu, 2016). The main application area of sentiment analysis is user generated content on the web like social media (Hutto and Gilbert, 2014; Schmidt et al., 2020b) or movie reviews (Kennedy and Inkpen, 2006). Next to sophisticated machine learning approaches, there are also rule-based approaches working with lexical resources and simple rules (Taboada et al., 2011; Schmidt and Burghardt, 2018). Unsurprisingly, sentiment analysis is a popular method to explore depression and social media. Wang et al. (2013) utilize lexicon-based sentiment analysis to calculate the chance for depression on micro-blogs. IBirjali et al. (2017) predict suicidal ideation in Twitter data via machine learning based sentiment analysis. Davcheva et al. (2019) explore three English-language mental health forums via sentiment analysis. Sentiment scores of users develop depending on several conditions (e.g. how active the user is).

To analyze linguistic and semantic word categories, we use the Linguistic Inquiry and Word Count (LIWC) dictionary (Pennebaker et al., 2015). It consists of multiple linguistic categories like 1st person singular, pronouns or adjectives as well as rather semantic and psychological categories like anxiety, female/male language or spirituality and a corresponding list of words that have been demonstrated to be connoted with this category. LIWC is an established and trusted resource in psychological investigations (Lee et al., 2015) but is also applied in areas outside of psychology (Schmidt et al., 2020a). Findings with LIWC include that people with depression use first person singular and negatively biased words more frequently than a control group (Lee et al., 2007).

Topic modeling is a method to create topical categories based on text documents without a priori subject definitions (Jockers, 2013). We apply Latent Dirichlet Allocation (LDA) topic modeling (Blei et al., 2003), which is one of the most established topic modeling approaches (Jockers, 2013). A topic is a list of words that frequently occur with each other in a set of documents. Given a number of expected topics, a LDA model produces lists of such words as a result. LDA models have also already been utilized to examine the use of language of depressive individuals. For example, Resnik et al. (2015) examined the use of supervised topic models in the analysis of linguistic signals for detecting depression.

There is numerous research examining depression and depression communities in social media cf. (Conway and O'Connor, 2016). De Choudhury and De (2014) apply various techniques and examine multiple mental health communities on Reddit. Among other, they characterize mental health social support into the categories "emotional", "informational", "prescriptive" and "instrumental", of which "emotional" and "prescriptive" are more likely to occur. De Choudhury and Kiciman (2017) continued their work by focusing on how the language of comments influences risk to suicidal ideation. Park and Conway (2017) use the LIWC dictionary to analyze r/depression and other health related forums to investigate user based longitudinal changes. They show that users with a long-term participation shifted to the use of more positive language and indicated that this leads to positive effects. Fraga et al. (2018) explore multiple mental health subreddits via discourse pattern analysis and found that the longest discussions are initiated by threads asking for help and that encouragement words are a frequent pattern.

3 Methods and Results

3.1 Data Acquisition and Corpora

For more information and access to parts of the used corpus, visit
`https://github.com/lauchblatt/OnlineDepressionForumsTextMining`

3.1.1 Subreddit r/depression

r/depression[2] is an English-speaking subreddit and consist of *submissions*, also called *threads* in other forums. Such submissions include a title and the initial content written by the author. Answers to a submission are called *comments*. Scraping of /r/depression was done using the Pushshift.io API Wrapper

---

[2]https://www.reddit.com/r/depression/

Metric	Beyond Blue	r/depression
Threads/Submissions	3,922	131,073
Answers/Comments	24,500	715,128
Posts (Threads+ Answers)	28,422	846,201
Tokens	4,982,391	60,632,208
Tokens per post	175.3	71.7
Tokens per initial post	229.2	186.0
Tokens per answers	166.7	50.7
Tokens per sentence	12.6	11.3
Comment/submission ratio	6.2	5.5

Table 1: Corpus Metrics

psaw [3] and the Python Reddit API Wrapper *praw* [4] .

1,007,134 posts (submissions and comments) from all submissions created in 2019 were gathered between 02.04.2020 and 02.20.2020. Due to the much smaller corpus available for Beyond Blue, we decided to limit the extraction to a single year, to keep the corpora comparable to some extent but still decided to include an entire year to avoid influences due to seasonal changes. 2,295 posts were filtered out due to lack of content, like empty posts or ones that only contain a link. Another 158,638 posts were not considered because they were already deleted by the author himself, moderators or spam filter. This leads to 846,201 posts consisting of 131,073 submissions (15.5%) and 715,128 comments (85.5%).

3.1.2 Beyond Blue

"Beyond Blue" is a non-profit organization funded by Australian governments and states (Jorm et al., 2005) and offers an English-speaking forum specifically for depression[5]. Beyond Blue offers support ranging from their website featuring many up to date articles on mental wellbeing and a welcoming atmosphere, to a 24-hour hotline, online chat, several different well-tailored forums and special information on current topics. We acquired permission by the Beyond Blue research team to scrape the content of their forum.

Beyond Blue was scraped using Python with the standard libraries *lxml* and *urllib*. The basic forum architecture is similar to Reddit, however, submissions are called *threads* and *posts* (equivalent to Reddit's comments) and cannot be nested, but may have a note referring to which post they are an answer. The only metadata available in the forum are the threads title and, respectively for every post, text, date and user information. Opposing to r/depression, we gathered all available threads and metadata in Beyond Blue, which still results in a noticeably smaller corpus, that contains posts from April 3rd 2013 until January 12th 2020. This results in 28,422 posts, of which 3,922 are initial posts (13.8%) of threads and 24,500 are answers inside threads (86.2%).

3.2 General Corpus Analysis

We used *SpaCy*[6] as central tool for the general corpus analysis. Table 1 illustrates general corpus statistics.

The Beyond Blue corpus consists of 4,982,391 tokens divided among 395,544 sentences. Considering the Beyond Blue corpus contains a total of 28,422 posts, this results in 175.3 tokens per post on average. There is a notable difference between the average token counts of initial posts of a thread (229.2) and answers (166.7). The average token count of a sentence is 12.6. Combining these results with the distribution of all posts into submissions and comments, the submissions (13.8% of all posts) are responsible for 18.0% of the tokens. r/depression contains a total of 60,632,208 tokens within 5,369,000 sentences.

---

[3]https://github.com/dmarx/psaw
[4]https://github.com/praw-dev/praw
[5]https://www.beyondblue.org.au/get-support/online-forums/depression
[6]https://spacy.io/

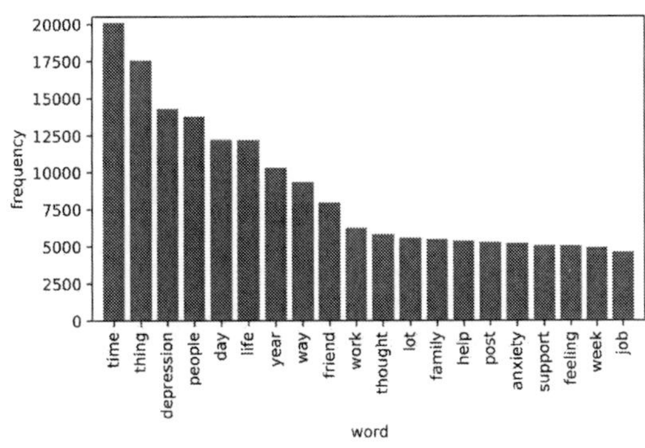

Figure 1: Twenty most frequent words of Beyond Blue

A post on r/depression contains 71.7 tokens on average. Splitting the posts into submissions and comments, the average token count is 186.0 for a submission and 50.7 for a comment. An average sentence contains 11.3 tokens. The submissions of r/depression, which make up 15.5% of all posts, contain 40.2% of all tokens.

Looking at the amount of submissions and comments, the comment/submission-ratios of 6.2 (Beyond Blue) and 5.5 (r/depression) are similar. The length of the posts makes the difference. By average, a post on Beyond Blue contains 144.5% more relevant tokens than a post on r/depression. The reason for this is that there are a lot of very short comments on r/depression as one can see by the very large difference concerning the tokens per answers. In addition, comparing the average token count of submissions, the token count on Beyond Blue is still a notable 20% larger than that of r/depression. These results lead to the assumption that users and moderators engage in much more length into the problems of initial posts.

3.3 Word Frequencies

We analyzed the most frequent words to gain a first impression of the topics and sentiment of the forums. Therefore, all stop words and tokens, which were not tagged as a noun, were removed from the corpora. We have chosen only nouns instead of all words, as these provide a better first overview of the corpora. The tokens were lemmatized and then counted to calculate the word frequency.

The most used word of the Beyond Blue corpus (figure 1) is "time" with 20,101 mentions. This is due to the users writing regarding to a specific time in their or other people's life. It is striking that the majority of the most frequent nouns are indeed other time related terms: "day" (0.245%), "year" (0.207%) and "week" (0.098%). "Depression", which is the main topic of the forum, is unsurprisingly the third most used word with 14,307 mentions (0.287%). Other frequent words like "friend" and "family", "work" and "job", "feeling" and "thought" or "help" and "support" possibly indicate some topics, which are explored in detail in the topic modeling section. Looking at the r/depression corpus (Figure 2), "time" is also the most used word (0.376%). The time-related terms are very common as well with 0.262% (day), 0.228% (year), 0.084% (month) and 0.080% (week), which could indicate that much of the conversation relies on narrating own experiences. In direct comparison, the top twenty most frequent words of both forums are quite similar. Four of the twenty most frequent words of Beyond Blue are not included in the ones of r/depression ("lot", "post", "anxiety", "support") and vice versa ("person", "school", "shit", "month"). The occurrence of "school" (0.107%) among the most frequent nouns might indicate a younger user base for r/depression.

3.4 Word Categories

To analyze the language used in both forums we split their content into language categories with the help of the LIWC dictionary. Results are received by assigning words of a given text to the linguistic and semantic word categories of the LIWC dictionary and yield the occurrence of each category in percentages as a result. Table 2 illustrates the results.

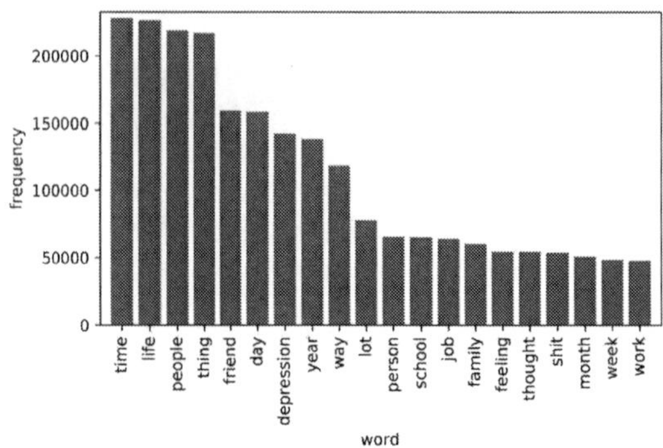

Figure 2: Twenty most frequent words of r/depression

LIWC Category	Examples for words in this category	Reddit Value	BB Value	Difference
Personal Pronoun	e.g. he, she, you	15.77	16.21	0.44
1st Person Singular	e.g. I, me, myself	15.01	12.78	2.23
1st Person Plural	e.g. our, we, us	0.99	1.45	0.46
2nd Person	e.g. y'all, you, yourself	4.34	5.37	1.03
3rd Person Singular	e.g. he, she, oneself	1.52	1.67	0.15
3rd Person Plural	e.g. them, they, they'll	0.78	0.66	0.12
Affective Processes	e.g. emotion, hopes, ugh and all terms of positive/negative emotion	14.30	14.72	0.42
Positive Emotion	e.g. better, fabulous, joy	4.43	4.81	0.38
Negative Emotion	e.g. bad, rotten, upset	7.12	6.54	0.58
Past Focus	e.g. ago, previous, remembered and other verbs in past tense	3.90	3.66	0.24
Present Focus	e.g. current, nowadays, understands and other verbs in present tense	17.70	17.95	0.25
Future Focus	e.g. expect, hopeful, wishing	1.60	1.65	0.05
Death	e.g. bury, die, kill	0.38	0.15	0.23
Swear Words	e.g. damn, hell, moron	1.33	0.74	0.59

Table 2: Percentages for LIWC word categories occurrence in r/depression and Beyond Blue

Sentiment Class	Beyond Blue	r/depression
Positive	40.20%	34.49%
Neutral	34.42%	34.20%
Negative	25.38%	31.32%

Table 3: Overall ratio of sentences classified with a polarity class

Sentiment Class	Beyond Blue		r/depression	
	initial	answer	initial	answer
Positive	29.12%	42.54%	28.44%	38.23%
Neutral	33.45%	34.62%	34.33%	34.12%
Negative	37.43%	22.84%	37.23%	27.65%

Table 4: Ratio of sentences classified with a polarity class, split by initial posts and answers

For clearer visualization, we only included a few categories, which either have large differences or are important for our context of depression forums. First, users of r/depression make more frequent use of negative language than those of Beyond Blue as shown by the *positive emotions* and *negative emotions* categories. This validates similar findings for sentiment analysis. Differences can also be seen in the rather negative categories *swear words* and *death*, where r/depression has nearly twice or three times the amount of Beyond Blue. Especially the difference considering swear words might be another indicator for a younger user base using more informal language. r/depression has a stronger *past focus* while Beyond Blue has a slightly stronger *present focus*. While both forums use about the same amount of pronouns, there is a notable difference in which pronouns they use. Within r/depression, the first person singular is more frequently used. Contrary, inside Beyond Blue first person plural and second person are more common. This might point to an important difference in communication since users in r/depression tend to talk more about themselves while the user base in Beyond Blue engages more in discussion directed towards each other. This is to some extent also in line with the finding that answers to initial posts are longer in Beyond Blue than in r/depression. Nevertheless, the results overall underline related findings that language of depressive individuals is negatively charged and includes frequent use of first person singular (Lee et al., 2007). Beyond Blue's smaller percentages in those two categories therefore imply that their user base utilizes less depressive language.

3.5 Sentiment Analysis

Sentiment analysis was conducted by using VADER (Hutto and Gilbert, 2014). VADER is a lexicon-based sentiment analysis tool that is specifically attuned to sentiments expressed in social media and is therefore also used for Reddit (Schmidt et al., 2020b). VADER shows good evaluation results for thie context of social media (Hutto and Gilbert, 2014). Therefore, it is a fitting selection for our sentiment approach. VADER provides sentiment analysis on a sentence-level. VADER classifies the given text with one of three polarities: positive, neutral and negative.

There is nearly the same ratio of neutral sentences in both forums, but positive and negative sentence ratios differ quite notably (see table 3). Beyond Blue consists of 40.20% positive sentences and 25.38% negative ones, whereas r/depression shows 34.49% positive and 31.32% negative sentences. Accordingly, r/depression generally has a more negative sentiment than Beyond Blue. The results of sentiment analysis when splitting initial posts and answers into sentences, are shown in Table 4.

The sentiment of initial posts is nearly identical for both forums. The answers are more positive compared to the initial posts overall, but answers on Beyond Blue are more positive than the ones on r/depression. According to the sentiment, the content that users post in one of these forums is approximately the same. The difference are the replies to those posts, which are more positive on Beyond Blue. This implies that the community in Beyond Blue reacts more supportive to people talking about their depression.

3.6 Topic Modeling

For our analysis, we created a Latent Dirichlet Allocation (LDA) model with the Python library *gensim* (Rehurek and Sojka, 2010) for each forum. We decided on creating one text document per post as it led to better results than using an entire thread. Tokenization and lemmatization were done for each post and stop words were removed. Posts which then had less than five tokens left were excluded from the topic modeling process, as they most likely will not contain any useful information. Originally, we created models with thirty topics, each represented by thirty keywords for both forums. Manual decrease for the number of topics to fifteen and the number of words to twenty lead to clearer visualization. The topics removed were mostly duplicates or topics with unclear results, the filtered keywords were duplicates and those which did not provide additional information. After testing with different variables, we decided on a chunk size of 2500 posts and 15 passes for each model, which generated meaningful topics.

Please refer to the appendix for the reduced and filtered 15 topics per forum and the corresponding 20 key terms for each topic (Appendix: Figure 3-4). Please note that the names for the topics are added by us based on our interpretation of the term list as LDA topic modeling does not produce names for topics. We will use these names in the following sub chapter.

While r/depression has entire topics revolving around school and college life (topics 2 ["School"] and 10 ["College"]) pointing towards user groups consisting of teenagers and young adults, Beyond Blue has several topics dealing with subjects like finances or insurance (e.g. topics 12 ["Social Care"] and 14 ["Work"]). This becomes even clearer when looking at the topic "Family" for each forum (topic 4 for r/depression and topic 6 for Beyond Blue). While the r/depression topic seems to be mainly from the viewpoint of someone growing up, with keywords like *mom* or *dad*, the Beyond Blue topic seems to be from the viewpoint of a parent with keywords like *boy* or *girl*.

Other differences can be found in the user's interests and hobbies of each forum. r/depression contains topics revolving around "Entertainment Media" (topic 12) or "Social Life" with friends (topic 9), whereas Beyond Blue has topics about general "Lifestyle" (topic 13, keywords e.g. *exercise, book, walk*) and "Local Life" (topic 5, keywords e.g. *dog, shop, town, car*). Similar topics, that occur for both forums include "Relationships" (r/depression topic 8; Beyond Blue topic 4), medical aspects (r/depression topics 14 ["Therapy"] and 15 ["Treatments"]; Beyond Blue topic 2 ["Mental Condition"], 10 ["Treatments"] and 11 ["Therapy"]), "Work" (r/depression topic 13, Beyond Blue topic 14) or "Family". Topics about relationships and medication are important in both forums and seem to be a consistent discussion point.

While topics concerning self-harm and suicide can also be found in both forums (topic 1 ["Emotional Pain"] and 3 ["Self-Harm"] for r/depression and topic 1 ["Self-Harm"] for Beyond Blue), they are represented very differently. In r/depression with keywords like *hate, cry, tired, emptiness, suffering, meaningless* and *torture*, the topics seem to be about venting or describing one's feelings and emotions that are connected to thoughts about self-harm. Even though this can be found in some words of the Beyond Blue topic as well (e.g. *anger, invisible*), it seems the underlying motive is about supporting other members of the community who have suicidal thoughts, shown by words like *encourage, concern, support* or *relate*. Beyond Blue contains more words like *reach, seek, offer* and *help* (topic 8 ["BB Community"]) or *suggestion, advice* and *sharing* (topic 3 ["BB Forum"]) paired with terms describing the forum itself like *BB, community* or *reply* and *post* which indicates that people offer help or reach out for help from other members of the forum. While the r/depression topics also include some words like *encourage* (topic 14 ["Therapy"]) or *recommend* (topic 15 ["Treatments"]), their occurrences are rare. All in all, similar to the ones about self-harm, most of the r/depression topics rather seem to be about describing one's own feelings and experiences (e.g. topics 9 ["Social Life"], 10 ["College"] and 11 ["Future Expectations"]).

4 Limitations

We want to highlight some of the major limitations of our project. The striking problem is that both corpora differ largely from each other, predominantly considering the size. We tried to control this limitation to some extent by (1) focusing on normalized metrics for our methods and (2) including at least a year for r/depression to avoid seasonal changes and to reduce the size of r/depression. The demographic of the user base might differ as well. Beyond Blue is a predominantly Australian forum while r/depression is constituted by a more international user base. While statistics show that Reddit's user base is primarly male and young[7], certain results point to a more adult user base for Beyond Blue (althoug we have no concrete statistics). r/depression has also no moderation by trained professionals. We address further differences between the user base in the upcoming chapter. Another important limitation lies in the application of some of the methods. The results of the sentiment analysis have to be taken with caution. While VADER is optimized for social media content, we did not precisely evaluate the performance on these specific corpora. While the LIWC dictionary is an established resource for lexical analysis, it is also not specifically designed for social media content and might therefore lack some important terms or categories.

5 Discussion

Please not that due to the fact that we gathered publicly available data and we only deal with anonymized data, we did not need permission by the ethics commission of our institution at the time of the writing.

First, we want to highlight similarities between both forums, which might point to general attributes of depression forums and their users: In both forums time-related terms are among the most frequent ones, showing that the narration of past and present events is a consistent discussion in depression forums. Both forums show a tendency to positive sentiment, thus illustrating that support and the recollection of positive aspects are an important part of depression forums in general. Topics about relationships, family, work and medication are highly important in both forums and structured rather similar. Especially the consistency considering the topics about relationships and family show the importance of these aspects for depressive illnesses.

Considering differences of the two forums, our findings correlate well and support assumptions over all applied methods. General corpus analysis and analysis about word categories conducted with LIWC show that the posts in general, as well as the language in Beyond Blue revolve around support and positivity, whereas negative speech and posts with little information value are more common in r/depression. This is in line with the finding that the answers to posts are much longer on average in Beyond Blue than in r/depression pointing to a more seriously engaged user and moderator base. The sentiment scores support this impression with Beyond Blue being noticeably more positive, mainly because answers to initial submissions in r/depression have a more negative sentiment than the rather positive responses on Beyond Blue. The findings produced by the topic modeling approach include the impression that the user base of Beyond Blue is older, as some topics have a focus on financial and organizational themes, and indicate a "parent view" while r/depression's users speak about school, college and their parents. Most interestingly, and complementing other findings, topics dealing with self-harm and suicide can be found in both forums but are considerably more support oriented in Beyond Blue and more emotional and negatively biased in r/depression.

Concluding, Beyond Blue and r/depression seem to have different focal points. Where Beyond Blue's support oriented language and content seems to indicate conversations where concrete problems are solved, r/depression's emotional posts, past tense and rather negative sentiment imply the focus lies on sharing experiences and converse about emotions. While our findings are exploratory at the moment, we hypothesize that a professionally curated and smaller forum might be more beneficial to be used by persons affected by depression since it is more supportive and positive in its content. This conclusion has to be taken with caution because (1) many differences might rely on the different user demographic, which seems to be younger in r/depression and (2) the open and more peer-oriented culture of Reddit might still be helpful and beneficial in some situations.

There are certain approaches we want to pursue to further investigate our results. We want to create a more balanced corpus by acquiring a collection of similar forums like Beyond Blue instead of just one. It is also necessary to optimize several of our text mining techniques for our context. Lastly, we also plan to work closely with professional psychologists and therapists to integrate their expertise into our project.

---

[7]https://www.techjunkie.com/demographics-reddit/

References

Marouane Birjali, Abderrahim Beni-Hssane, and Mohammed Erritali. 2017. Machine learning and semantic sentiment analysis based algorithms for suicide sentiment prediction in social networks. *Procedia Computer Science*, 113:65–72.

David M Blei, Andrew Y Ng, and Michael I Jordan. 2003. Latent dirichlet allocation. *Journal of machine Learning research*, 3(Jan):993–1022.

Mike Conway and Daniel O'Connor. 2016. Social media, big data, and mental health: current advances and ethical implications. *Current opinion in psychology*, 9:77–82.

Elena Davcheva, Martin Adam, and Alexander Benlian. 2019. User dynamics in mental health forums–a sentiment analysis perspective.

Munmun De Choudhury and Sushovan De. 2014. Mental health discourse on reddit: Self-disclosure, social support, and anonymity. In *Eighth international AAAI conference on weblogs and social media*.

Munmun De Choudhury and Emre Kiciman. 2017. The language of social support in social media and its effect on suicidal ideation risk. In *Eleventh International AAAI Conference on Web and Social Media*.

Barbara Silveira Fraga, Ana Paula Couto da Silva, and Fabricio Murai. 2018. Online social networks in health care: A study of mental disorders on reddit. In *2018 IEEE/WIC/ACM International Conference on Web Intelligence (WI)*, pages 568–573. IEEE.

Clayton J Hutto and Eric Gilbert. 2014. Vader: A parsimonious rule-based model for sentiment analysis of social media text. In *Eighth international AAAI conference on weblogs and social media*.

Matthew L Jockers. 2013. *Macroanalysis: Digital methods and literary history*. University of Illinois Press.

Anthony F Jorm, Anthony F Jorm, Helen Christensen, Kathleen M Griffiths, Anthony F Jorm, Helen Christensen, and Kathleen M Griffiths. 2005. The impact of beyondblue: the national depression initiative on the australian public's recognition of depression and beliefs about treatments. *Australian & New Zealand Journal of Psychiatry*, 39(4):248–254.

Alistair Kennedy and Diana Inkpen. 2006. Sentiment classification of movie reviews using contextual valence shifters. *Computational intelligence*, 22(2):110–125.

Chang H Lee, Myungju Lee, Sungwoo Ahn, and Kyungil Kim. 2007. Preliminary analysis of language styles in a sample of schizophrenics. *Psychological reports*, 101(2):392–394.

Changhwan Lee, Kyungil Kim, Jeongsub Lim, and Yoonhyoung Lee. 2015. Psychological research using linguistic inquiry and word count (liwc) and korean linguistic inquiry and word count (kliwc) language analysis methodologies. *Journal of Cognitive Science*, 16(2):132–49.

Bing Liu. 2016. *Sentiment analysis: Mining opinions, sentiments, and emotions*. Cambridge University Press.

Albert Park and Mike Conway. 2017. Longitudinal changes in psychological states in online health community members: understanding the long-term effects of participating in an online depression community. *Journal of medical Internet research*, 19(3):e71.

James W Pennebaker, Ryan L Boyd, Kayla Jordan, and Kate Blackburn. 2015. The development and psychometric properties of liwc2015. Technical report.

Radim Rehurek and Petr Sojka. 2010. Software framework for topic modelling with large corpora. In *In Proceedings of the LREC 2010 Workshop on New Challenges for NLP Frameworks*. Citeseer.

Philip Resnik, William Armstrong, Leonardo Claudino, Thang Nguyen, Viet-An Nguyen, and Jordan Boyd-Graber. 2015. Beyond lda: exploring supervised topic modeling for depression-related language in twitter. In *Proceedings of the 2nd Workshop on Computational Linguistics and Clinical Psychology: From Linguistic Signal to Clinical Reality*, pages 99–107.

Thomas Schmidt and Manuel Burghardt. 2018. An evaluation of lexicon-based sentiment analysis techniques for the plays of gotthold ephraim lessing. In *Proceedings of the Second Joint SIGHUM Workshop on Computational Linguistics for Cultural Heritage, Social Sciences, Humanities and Literature*, pages 139–149.

Thomas Schmidt, Isabella Engl, Juliane Herzog, and Lisa Judisch. 2020a. Towards an analysis of gender in video game culture: Exploring gender-specific vocabulary in video game magazines. In *Proceedings of the Digital Humanities in the Nordic Countries 5th Conference (DHN 2020*, pages 333–341.

Thomas Schmidt, Florian Kaindl, and Christian Wolff. 2020b. Distant reading of religious online communities: A case study for three religious forums on reddit. In *Proceedings of the Digital Humanities in the Nordic Countries 5th Conference (DHN 2020*, pages 157–172.

Maite Taboada, Julian Brooke, Milan Tofiloski, Kimberly Voll, and Manfred Stede. 2011. Lexicon-based methods for sentiment analysis. *Computational linguistics*, 37(2):267–307.

Xinyu Wang, Chunhong Zhang, Yang Ji, Li Sun, Leijia Wu, and Zhana Bao. 2013. A depression detection model based on sentiment analysis in micro-blog social network. In *Pacific-Asia Conference on Knowledge Discovery and Data Mining*, pages 201–213. Springer.

A   Appendix

	1	2	3	4	5	6	7	8	9	10	11	12	13	14	15
	Emotional Pain	School	Self-Harm	Family	Money	Physical Condition	Mental Condition	Relationships	Social Life	College	Future Expectations	Entertainment Media	Work	Therapy	Treatments
1	love	school	live	parent	money	brain	mental	relationship	friend	college	future	enjoy	work	help	doctor
2	shit	teacher	die	mom	pay	effect	issue	guy	family	fail	happiness	watch	job	need	medication
3	hate	bully	suicide	dad	buy	healthy	health	girl	close	class	dream	game	quit	therapist	med
4	fuck	counselor	death	mother	save	weight	illness	girlfriend	social	study	goal	fun	career	therapy	psychiatrist
5	fucking	black	alive	brother	afford	exercise	physical	date	meet	grade	reality	music	company	listen	hospital
6	hurt	winter	attempt	sister	poor	smoke	panic	boyfriend	lonely	graduate	society	video	apply	advice	antidepressant
7	kill	ghost	option	father	bill	weed	chemical	woman	group	test	purpose	interest	position	support	treatment
8	cry	white	decision	cat	cost	gym	patient	ex	conversation	student	joy	hobby	fire	reach	pill
9	pain	lunch	choice	son	debt	gain	chronic	wife	online	university	create	movie	area	professional	recommend
10	tired	senior	survive	animal	rent	habit	ADHD	abuse	message	semester	opportunity	medium	project	offer	episode
11	cut	bus	existence	argument	law	mechanism	Wellbutrin	young	friendship	exam	achieve	art	program	suggest	appointment
12	lie	military	suffering	sadly	education	addiction	imbalance	partner	party	final	path	tv	business	helpful	medical
13	stupid	summer	meaningless	replace	motivate	lift	address	sex	loneliness	disappoint	success	favorite	financial	willing	switch
14	hell	classmate	cruel	threaten	expensive	unhealthy	OCD	adult	connection	highschool	desire	boring	boss	psychologist	prescribe
15	heart	popular	torture	aunt	interview	meditation	psychiatric	baby	isolate	homework	ahead	YouTube	store	community	risk
16	emptiness	beer	workout	ease	earn	body	psychological	husband	join	major	successful	bored	office	provide	insurance
17	inside	transfer	fighting	couch	dollar	motivated	eating	divorce	invite	academic	expectation	computer	shift	session	psych
18	sick	September	inevitable	grandma	insurance	active	disagree	marry	distance	course	direction	distract	field	resource	available
19	exist	ditch	empathize	insult	car	balance	apathy	raise	interaction	score	succeed	series	coworker	encourage	dangerous
20	dead	withdraw	sorrow	adopt	unmotivated	lifestyle	conflict	cheat	circle	freshman	pursue	passionate	team	counsellor	withdrawal

Figure 3: Cleaned LDA topics for r/depression

	1	2	3	4	5	6	7	8	9	10	11	12	13	14	15
	Self-Harm	Mental Condition	BB Forum	Relationships	Local Life	Family	Alcoholism	BB Community	Physical Health/ Food	Treatments	Therapy	Social Care	Lifestyle	Work	Emotional/Weather
1	suicide	mental	thank	love	dog	child	alcohol	help	eat	medication	psych	team	find	job	cry
2	harm	health	reply	life	group	mum	drinking	support	weight	mood	psychologist	claim	work	work	tear
3	self	illness	post	relationship	cat	parent	drug	forum	food	bipolar	appointment	ps	exercise	money	emotion
4	phone	depression	thread	believe	town	husband	alcoholic	welcome	ocd	anti	go	private	mind	pay	song
5	comment	anxiety	read	abuse	live	daughter	drunk	professional	healthy	depressant	psychiatrist	Psychologist	book	study	cold
6	encourage	symptom	response	woman	car	son	sober	find	binge	effect	session	charge	good	uni	hug
7	community	issue	appreciate	hurt	city	mother	bottle	seek	gain	diagnosis	doctor	disability	enjoy	boss	weather
8	concern	suffer	share	self	house	old	loathing	post	eating	doctor	see	centrelink	walk	financial	fight
9	service	condition	advice	soul	pet	young	mate	go	disorder	psychiatrist	visit	clinic	positive	apply	crying
10	moderator	treatment	story	beautiful	country	family	booze	counsellor	body	episode	refer	fee	help	career	rain
11	general	experience	kind	forgive	local	sister	stop	advice	kg	disorder	plan	hr	read	stress	sun
12	supportive	professional	respond	human	animal	dad	frighten	BB	habit	symptom	therapy	cover	thought	employment	flow
13	invisible	cause	comment	happiness	club	wife	Alcohol	reach	intrusive	prescribe	treatment	safety	day	quit	burst
14	issue	chronic	helpful	pain	meet	home	meeting	offer	loss	depressive	discuss	department	goal	interview	shine
15	relate	medical	forum	special	move	brother	sincerely	helpful	genetic	ad	hospital	insurance	activity	position	warm
16	support	test	lovely	way	home	baby	tragedy	hope	label	take	patient	worker	learn	bill	emotional
17	clinically	Depression	encouragement	live	shop	father	licence	good	perfectionist	therapy	therapist	system	time	debt	dark
18	anger	admission	suggestion	laugh	walk	love	merry	chat	meal	dosage	Medicare	emergency	focus	workplace	inside
19	suicidal	PTSD	supportive	strength	volunteer	girl	poison	community	appetite	change	medical	payment	achieve	company	ocean
20	attempt	recover	sharing	deserve	area	boy	drink	suggest	product	BPD	recommend	policy	play	income	bright

Figure 4: Cleaned LDA topics for Beyond Blue

Towards Preemptive Detection
of Depression and Anxiety in Twitter

David Owen   **Jose Camacho-Collados**   **Luis Espinosa-Anke**

School of Computer Science and Informatics

Cardiff University, United Kingdom

`{owendw1,camachocolladosj,espinosa-ankel}@cardiff.ac.uk`

Abstract

Depression and anxiety are psychiatric disorders that are observed in many areas of everyday life. For example, these disorders manifest themselves somewhat frequently in texts written by non-diagnosed users in social media. However, detecting users with these conditions is not a straight-forward task as they may not explicitly talk about their mental state, and if they do, contextual cues such as immediacy must be taken into account. When available, linguistic flags pointing to probable anxiety or depression could be used by medical experts to write better guidelines and treatments. In this paper, we develop a dataset designed to foster research in depression and anxiety detection in Twitter, framing the detection task as a binary tweet classification problem. We then apply state-of-the-art classification models to this dataset, providing a competitive set of baselines alongside qualitative error analysis. Our results show that language models perform reasonably well, and better than more traditional baselines. Nonetheless, there is clear room for improvement, particularly with unbalanced training sets and in cases where seemingly obvious linguistic cues (keywords) are used counter-intuitively.

1 Introduction

Mental illnesses are psychiatric disorders that may cause sufferers significant distress and impair their ability to function in social and work activities (Bolton, 2008). The most prevalent mental illnesses are depression and anxiety, which are estimated to affect nearly one in ten people worldwide (676 million) according to a recent study (World Health Organization, 2016). While depression and anxiety are different disorders, they also share symptoms and, thus, clinicians often diagnose patients with both illnesses at consultation.[1] Identifying these conditions at early stages is relevant not only because of their inherent importance, but also because they are precursors to major related concerns in public health including self-harm (Centers for Disease Control and Prevention, 2015), making timely diagnosis and treatment even more essential. However, sufferers of depression and anxiety can find that it takes great courage and strength to seek professional treatment (Dennis C Miller, 2016). They may also be afraid to confide in their peers due to the stigma of mental illness (Wasserman et al., 2012).

With reluctance to seek professional treatment or rely on their peers, sufferers often turn to online resources for support. These include both specialised and general communities, with Twitter and Reddit (Yates et al., 2017) being paramount examples of the latter. Because of this, systems that can automatically detect and flag such cases at a large-scale are highly desirable (Guntuku et al., 2017). They may enable prompt analysis and treatment, which is crucial in the early development of such conditions. Moreover, the interpersonal and economic effects of these illnesses may be mitigated with prompt intervention (Lexis et al., 2011).

In this paper, we build a classification dataset[2] to assist in the detection of depression and anxiety in Twitter, and compare several text classification baselines. The results show that state-of-the-art language models (LMs henceforth) like BERT (Devlin et al., 2019) unsurprisingly outperform competing

---

[1]https://www.bupa.co.uk/newsroom/ourviews/2017/10/anxiety-depression

[2]The datasets and code used in our experiments are available online at the following repository: `https://bitbucket.org/nlpcardiff/preemptive-depression-anxiety-twitter`.

Proceedings of the 5th Social Media Mining for Health Applications (#SMM4H) Workshop & Shared Task, pages 82–89

Barcelona, Spain (Online), December 12, 2020.

baselines. However, when the dataset shows an unbalanced distribution, linear models perform on par. Finally, alongside quantitative results, we also provide a qualitative analysis through which we aim to better understand the strengths and limitations of the models under study. Further, we identify the linguistic patterns alluding to the presence of depression and anxiety that elude all of the classifiers, and consider how we might improve performance against such patterns in the future.

2   Related Work

Baclic et al. (2020) surveyed the use of NLP in healthcare and identified the inherent opportunities and challenges. NLP permits speedy analysis of large volumes of unstructured text such as electronic patient records or social media posts, which can help support early healthcare interventions. For example, automated analysis of consumers' online reviews was used to predict the presence of depression in reviewers prior to its formal diagnosis (Harris et al., 2014). However, Baclic et al. noted that the effectiveness of NLP in the domain of mental health is constrained by a lack of high quality training data. Consequently, we chose to build our own labelled dataset.

In terms of detecting depression and anxiety in social media, the work of Yates et al. (2017) is perhaps the most related to ours. They showed that depression among Reddit participants can be detected by identifying certain lexical and psycholinguistic features in the contents of their Reddit postings. These features include indications of negative psychological processes such as anger and sadness, which may be denoted by the words "hate" and "grief" respectively (Pennebaker et al., 2015). The work of Yates et al. is therefore highly relevant to ours since it concerns automated detection of mental illness in written social media discourse. However, there are three key aspects that set our paper apart, namely: (1) We identify users of social media platforms who appear to have *not* yet been diagnosed with mental illness. We do this in the spirit of providing the opportunity for early healthcare intervention in these cases; (2) we consider automated classification using feature rich representations of users' online postings using state-of-the-art NLP methods such as word embeddings and pre-trained LMs; and (3) we consider concise written discourse in the form of tweets, rather than Reddit postings, which are generally more verbose.

3   Dataset Construction

In this section, we describe our process for building a dataset for detecting depression and anxiety in Twitter. We describe the tweet collection procedure (Section 3.1), annotation (Section 3.2), and provide information about the inter-annotator agreement (Section 3.3).

3.1   Tweet collection

First, we used Twitter's Stream API to compile a large corpus of tweets. All tweets were of English language and published between May 2018 and August 2019.[3] We only considered tweets containing at least three tokens and without URLs so as to avoid bot tweets and spam advertising. All personal information, including usernames (denoting the author or other users) and location were removed from the corpus - only textual information was retained. We did however retain emojis and emoticons. We surmised that they may, in part at least, be indicative of depression or anxiety.

The corpus was then filtered. We aimed to identify tweets whose authors may be suffering from depression or anxiety but may not yet have been diagnosed by a clinician. To achieve this, we sought tweets containing occurrences of *depress*, *anxie*, or *anxio*, but not *diagnos*[4] - an approach similar to that used by Bathina et al. (2020). This produced an initial set of 89,192 tweets. From these tweets we proceeded to annotate a random subset of 1,050 tweets to arrive at our dataset.

3.2   Annotation

Three human annotators were appointed. The prerequisites for these annotators were to be fluent in English and to have familiarity with Twitter. The 1,050 tweet dataset was divided into three distinct subsets of 300 and one distinct shared subset of 150. Each annotator received one of the former subsets in addition to the latter subset. They were tasked with labelling the 450 tweets that they had received.

Annotators 1 and 2	Annotators 1 and 3	Annotators 2 and 3	Average Pairwise Agreement	Fleiss' Kappa	Krippendorff's Alpha
78.67	80.67	80.67	80.00	0.60	0.60

Table 1: Pairwise Percentage Agreements and Inter-Annotator reliability.

One of two numerals was selected by the annotator with respect to each tweet:

1: The tweeter appears to be suffering from depression or anxiety.

0: The tweeter does not appear to be suffering from depression or anxiety.

Guidelines were compiled to aid the annotation exercise. Their purpose was to ensure a consistent approach amongst annotators and to resolve ambiguous cases. These guidelines are defined below along with examples and their suggested labels:

1. The tweeter states that they have depression or anxiety
   Example: *"I feel sick to my stomach, I hate having such bad anxiety"* - **1**

2. The tweeter states that they have had depression or anxiety in the past
   Example: *"Counselling fixed my depression"* - **0**

3. The tweeter is referring to a fellow tweeter who may have depression or anxiety
   Example: *"@user I wish you all the best in beating your anxiety"* - **0**

4. The tweeter is temporarily depressed or anxious due to a short or superfluous event
   Example: *"Nothing gives me anxiety more than the tills at Aldi"* - **0**

5. The tweet is ambiguous or does not provide definitive information
   Example: *"Depression is not taken serious enough"* - **1**

Guideline 5 recommends positive labelling in ambiguous cases. This is to help achieve high recall in terms of tweeters who appear to be suffering from depression or anxiety. Whilst this approach will inevitably retrieve negative instances, an eventual real-world application would require all retrieved instances to be verified manually by medical experts, and therefore high recall at the expense of lower precision is an acceptable tradeoff. In fact, it is recommended that results from automatic classifiers used in healthcare settings should be verified via an "expert-in-the-loop approach" (Holzinger, 2016).

The guidelines evolved following the annotators' first attempts at the exercise. A conflict resolution meeting revealed that while agreement was due to be acceptable, there were instances where annotators felt unable to assign either label. This gave rise to the addition of guideline 5, which allowed the annotators to complete the exercise with confidence.

3.3 Inter-Annotator Agreement

Once the annotation was completed, we calculated the Average Pairwise Percentage Agreement of the three annotators with respect to the 150 common tweets that they had received (Table 1).

An average pairwise agreement of 80% was recorded. To validate the quality of the exercise two further measures of inter-annotator reliability were selected: Fleiss' Kappa and Krippendorf's Alpha. They are apt for inter-annotator exercises involving more than two annotators (Zapf et al., 2016). Both measures returned scores indicating "substantial agreement" amongst the annotators (Xie et al., 2017).

Confidence in the annotation guidelines was therefore established versus the 150 tweets common to each annotator. Disagreements in the labels were decided by majority voting among the three annotators. For example, the tweet *"My seasonal depression automatically begun tonight at 12am"* was labelled **1** by two of the annotators and labelled **0** by the third annotator. However, majority voting meant that it was finally labelled **1**. The annotators then proceeded to label their distinct 300 tweet subsets independently.

4 Experimental evaluation

In the following we detail the experimental setting (Section 4.1) and then present the results (Section 4.2) and an analysis (Section 4.3).

	Training		**Test**
	DATD	DATD+Rand	DATD
Positive Instances	473	473	75
Negative Instances	427	4,027	75
Total Instances	900	4,500	150

Table 2: Characteristics of the datasets used in the evaluation.

4.1 Experimental setting

4.1.1 Data

We prepared our annotated dataset described in Section 3 for input to a series of supervised classifiers. The three distinct subsets of 300 annotated tweets were combined to form a training set of 900 tweets. The 150 tweets labelled by all annotators formed the test set. We named this dataset *DATD* (Depression and Anxiety in Twitter Dataset). The test set's ratio of positive instances to negative instances was exactly 1:1 following the annotation exercise. This contrasts with related published datasets upon which no annotation had been performed and all instances were deemed mental illness-related.[5]

Following a similar approach to Bathina et al. (2020), we also compiled a non-annotated set of 3,600 random tweets which did not contain any occurrence of *depress*, *anxie*, *anxio*, or *diagnos*. These were merged with the 900 tweet training set to form a larger training set of 4,500 tweets. The purpose of this large training set (DATD+Rand henceforth) was to recreate a more realistic (and noisy) setting where most training instances are negative. This meant that only 10.5% of the instances in this training set contained any of the keywords used to compile the positive examples. The 150 tweets labelled by all annotators formed the test set once again. The main characteristics of the two datasets are summarised in Table 2.

4.1.2 Comparison systems

We evaluated several binary classifiers on both the DATD and DATD+Rand datasets, guided by existing research concerning problems similar to the one at hand. To this end, we deemed a Support Vector Machine (SVM) and an LM to be suitable classifiers. SVMs have demonstrated effectiveness when used with Twitter datasets in healthcare contexts (Prieto et al., 2014; Han et al., 2020). For our experiments we used both a standard SVM classifier with TF-IDF features and a classifier based on the average of word embeddings within the tweet.

With regards to pre-trained LMs, we used BERT (Bidirectional Encoder Representations from Transformers) and ALBERT (A Lite BERT) (Lan et al., 2019). These LMs have been deployed effectively in NLP tasks, leading to state-of-the-art results in most standard benchmarks (Wang et al., 2019) including Twitter (Basile et al., 2019; Roitero et al., 2020). In particular, ALBERT has been shown to provide competitive results despite being relatively light-weight compared to other LMs.

Finally, for completeness we added a naïve baseline that predicts positive instances in all cases.

4.1.3 Training details

We used the scikit-learn SVM model (Pedregosa et al., 2011) as well as its TF-IDF (Term Frequency-Inverse Document Frequency) vectorizer implementations.[6] The word embeddings generated for each tweet were drawn from vectors trained on Twitter data (Pennington et al., 2014, GloVe). These vectors had a dimensionality of 200, and so did the averaged embedding generated.

We performed tweet text preprocessing prior to their input to the SVM. In one series of SVM experiments all tweets underwent tokenization and lowercasing only, but in a second series all tweets also underwent tweet specific preprocessing[7] (SVM+preproc henceforth). The preprocessing entailed the removal of hashtags, user mentions, reserved words (such as "RT" and "FAV"), emojis, and smileys. This enabled us to see how the presence of these common tweet features affected classification performance. In both cases, the SVM used a linear kernel and default hyperparameters.

---

[5]https://github.com/AshwanthRamji/Depression-Sentiment-Analysis-with-Twitter-Data/blob/master/tweetdata.txt

[6]https://scikit-learn.org/stable/modules/generated/sklearn.feature_extraction.text.TfidfVectorizer.html

[7]https://pypi.org/project/tweet-preprocessor/

Classifier	Features	Accuracy	Precision	Recall	F1
SVM	TF-IDF	0.633	0.619	0.693	0.654
	Word Embs	0.727	0.693	0.813	0.748
	TF-IDF + Word Embs	0.733	0.711	0.787	0.747
SVM+preproc	TF-IDF	0.633	0.616	0.707	0.658
	Word Embs	0.713	0.695	0.760	0.726
	TF-IDF + Word Embs	0.727	0.698	0.800	0.745
BERT	LM	**0.749**	**0.713**	**0.856**	**0.774**
ALBERT	LM	0.675	0.651	0.779	0.705
Naïve baseline	-	0.500	0.500	1.000	0.670

Table 3: Results of the first experimental setup (DATD).

To deploy the LM classifiers we used the Simple Transformers[8] software library. It provides a convenient Application Programming Interface (API) to the Transformers Library, which itself provides access to BERT and ALBERT models, amongst others (Wolf et al., 2019). The BERT and ALBERT classifiers used were "bert-base-uncased" and "albert-base-v1",[9] respectively.

Unlike the SVM experiments it was not necessary to tokenize tweet texts prior to their input to the BERT or ALBERT classifiers; they perform their own tokenization. Tweet texts did undergo prior lowercasing however. The classifiers were instantiated with Simple Transformers' default hyperparameters.

4.2 Results

Experimental results are presented in Table 3 (DATD) and Table 4 (DATD+Rand).[10]

In the first setting, BERT achieves the best overall results, which is not unexpected. More importantly, the overall accuracy (i.e. 0.749) is close to the pairwise IAA (i.e. 0.800), which suggests that BERT is able to follow the guidelines provided in Section 3.2 to a reasonable extent. As for the linear models, the Twitter-specific preprocessing for the SVM does not lead to any improvements. In the following analysis section, we aim at shedding light on the types of error made by these models.

In the second setting where random tweets are added as negative instances in the training sets (Table 4), SVM with word embeddings features perform similarly to BERT, being in fact slightly better overall. This result is perhaps surprising, but may be due to the relative robustness of SVMs with respect to unbalanced training sets, which seem to have a greater effect on the LMs. Another explanation may be that by concatenating TF-IDF features and word embeddings the classifier is effectively leveraging both global and local dependencies, which have been shown to be crucial in tweet classification tasks such as emoji prediction (Barbieri et al., 2018) and stance detection (Mohammad et al., 2016).

More generally, the results in this setting are not hugely different from the first setting's. This is encouraging, as it suggests that supervised models can also perform in a more realistic setting where the negative instances are more prevalent than the positive ones.

4.3 Analysis

Perhaps the main highlights of our experiments are the results obtained by BERT and the concatenation of TF-IDF features and word embeddings in SVMs. BERT performs remarkably well despite not being trained on Twitter data. This could suggest that, although slang, jargon, misspellings, and emoji are typical in microposts, users suffering from mental illness are more articulate in their online writing than their mentally healthy counterparts. Thus their writing style is more likely to be picked up by an LM with restricted vocabularies.[11] Another surprising set of results concerns ALBERT, which was trained on the same corpus as BERT. We could expect that given the modest size of this dataset, an overparameterized

---

[8]https://github.com/ThilinaRajapakse/simpletransformers

[9]https://huggingface.co/transformers/pretrained_models.html

[10]Results for the LMs BERT and ALBERT were reported for the average of five different runs.

[11]While this is overcome within BERT by the use of WordPiece (Wu et al., 2016), the quality of its internal representations degrade as the number of OOV words it has to deal with increases.

Classifier	Features	Accuracy	Precision	Recall	F1
	TF-IDF	0.673	0.681	0.653	0.667
SVM	Word Embs	0.740	0.737	0.747	0.742
	TF-IDF + Word Embs	**0.747**	**0.740**	0.760	**0.750**
	TF-IDF	0.660	0.658	0.667	0.662
SVM+preproc	Word Embs	0.720	0.704	0.760	0.731
	TF-IDF + Word Embs	0.740	0.725	0.773	0.748
BERT	LM	0.693	0.656	0.851	0.737
ALBERT	LM	0.648	0.609	**0.880**	0.715
Naïve baseline	-	0.500	0.500	1.000	0.670

Table 4: Results of the second experimental setup (DATD+Rand).

model like BERT could fall short when compared to lighter versions, but this does not seem to be the case. In any case, ALBERT achieves the highest recall score on the DATD+Rand dataset.

Let us now highlight a selection of tweets in the DATD dataset and the performance of the classifier configurations against them. For example, the tweet *"got a yellow phone case hoping it will cure my depression"* was labelled **1**, a label only predicted by two of the eight configurations, namely BERT and SVM with TF-IDF + Word Embs features. This tweet is a good example of the overarching complexity of the problem, the presence of prosaic terms like "phone case", or positive words like "hope" or "cure" may have confused the simpler word-based models.

Another illustrative example is *"you know that i'm the best, is that why you depressed?"*, which was labelled **0** and was misclassified by all configurations. We hypothesize that this may be due to having three instances of two distinct pronouns ("I" and "you"), which are likely used often by depressed or anxious tweeters, although it was not the case in this particular example.

There are other interesting examples, including cases where only the LMs (BERT and ALBERT) made correct predictions. For example, *"Got my first call center job and my anxiety is through the roof"*, which was labelled **1**, and *"Hot shot screaming gives me anxiety watching these game lol mans be stressed but these games good as hell learning a lot."*, which was labelled **0**. Conversely, there are also cases where only SVMs made correct predictions. For example *"big ass spider in my room and it disappeared so I'll just have anxiety for the rest of the night I'm in here"*, which was labelled **0**. While it is not clear whether there is a systematic pattern to draw conclusions from, it does seem that when only the LMs succeed there is some degree of world or semantic understanding required to capture the condition of the tweeter (for example, the fact that you are probably not actually anxious by watching a game).

5   Conclusion and Future Work

In this paper, we have presented an experimental evaluation for detecting depression and anxiety in social media. We have developed a dataset, DATD, for predicting depression and anxiety in Twitter. Using this dataset we have run a comparative analysis of pre-trained LMs and traditional linear models. Not surprisingly, LMs performed relatively well on this task with a balanced set, but they do not outperform lighter-weight methods when the training data is unbalanced. Given the relatively small size of the dataset, we have also performed a qualitative analysis to identify areas for improvement.

Since these automatic models are intended for use by medical experts future work could involve collaboration with them. Moreover, classifier performance must be measured and certified versus large, heterogeneous datasets before adoption in healthcare is likely to be considered (Kelly et al., 2019). Collaboration may serve mental health experts to better understand social media at a large-scale, and to develop better guidelines and treatments.

Finally, we are also planning to extend this work to develop a dataset with finer grained distinctions, similar to Bathina et al. (2020) for Cognitive Distortion Schemas.

References

Oliver Baclic, Matthew Tunis, Kelsey Young, Coraline Doan, Howard Swerdfeger, and Justin Schonfeld. 2020. Challenges and opportunities for public health made possible by advances in natural language processing. *Canada Communicable Disease Report*, 46(6):161–168.

Francesco Barbieri, Jose Camacho-Collados, Francesco Ronzano, Luis Espinosa Anke, Miguel Ballesteros, Valerio Basile, Viviana Patti, and Horacio Saggion. 2018. Semeval 2018 task 2: Multilingual emoji prediction. In *Proceedings of The 12th International Workshop on Semantic Evaluation*, pages 24–33.

Valerio Basile, Cristina Bosco, Elisabetta Fersini, Nozza Debora, Viviana Patti, Francisco Manuel Rangel Pardo, Paolo Rosso, Manuela Sanguinetti, et al. 2019. Semeval-2019 task 5: Multilingual detection of hate speech against immigrants and women in twitter. In *13th International Workshop on Semantic Evaluation*, pages 54–63. Association for Computational Linguistics.

Krishna C. Bathina, Marijn ten Thij, Lorenzo Lorenzo-Luaces, Lauren A Rutter, and Johan Bollen. 2020. Depressed individuals express more distorted thinking on social media. *arXiv preprint arXiv:2002.02800*.

Derek Bolton. 2008. *What is mental disorder? : an essay in philosophy, science, and values*. International perspectives in philosophy and psychiatry. Oxford University Press, Oxford ; New York.

Centers for Disease Control and Prevention. 2015. Suicide: Facts at a glance [fact sheet].

Dennis C Miller. 2016. Mental health awareness month: Take the first step towards a mentally healthy workplace.

Jacob Devlin, Ming-Wei Chang, Kenton Lee, and Kristina Toutanova. 2019. Bert: Pre-training of deep bidirectional transformers for language understanding. In *Proceedings of the 2019 Conference of the North American Chapter of the Association for Computational Linguistics: Human Language Technologies, Volume 1 (Long and Short Papers)*, pages 4171–4186.

Sharath Chandra Guntuku, David B Yaden, Margaret L Kern, Lyle H Ungar, and Johannes C Eichstaedt. 2017. Detecting depression and mental illness on social media: an integrative review. *Current Opinion in Behavioral Sciences*, 18:43–49.

Kai-Xu Han, Wei Chien, Chien-Ching Chiu, and Yu-Ting Cheng. 2020. Application of support vector machine (svm) in the sentiment analysis of twitter dataset. *Applied Sciences*, 10(3):1125.

Jenine K Harris, Raed Mansour, Bechara Choucair, Joe Olson, Cory Nissen, and Jay Bhatt. 2014. Health department use of social media to identify foodborne illness—chicago, illinois, 2013–2014. *MMWR. Morbidity and mortality weekly report*, 63(32):681.

Andreas Holzinger. 2016. Interactive machine learning for health informatics: when do we need the human-in-the-loop? *Brain Informatics*, 3(2):119–131.

Christopher J Kelly, Alan Karthikesalingam, Mustafa Suleyman, Greg Corrado, and Dominic King. 2019. Key challenges for delivering clinical impact with artificial intelligence. *BMC medicine*, 17(1):195.

Zhenzhong Lan, Mingda Chen, Sebastian Goodman, Kevin Gimpel, Piyush Sharma, and Radu Soricut. 2019. Albert: A lite bert for self-supervised learning of language representations. *arXiv preprint arXiv:1909.11942*.

Monique AS Lexis, Nicole WH Jansen, Marcus JH Huibers, Ludovic GPM Van Amelsvoort, Ate Berkouwer, Gladys Tjin A Ton, Piet A Van Den Brandt, and IJmert Kant. 2011. Prevention of long-term sickness absence and major depression in high-risk employees: a randomised controlled trial. *Occupational and Environmental Medicine*, 68(6):400–407.

Saif Mohammad, Svetlana Kiritchenko, Parinaz Sobhani, Xiaodan Zhu, and Colin Cherry. 2016. Semeval-2016 task 6: Detecting stance in tweets. In *Proceedings of the 10th International Workshop on Semantic Evaluation (SemEval-2016)*, pages 31–41.

Fabian Pedregosa, Gaël Varoquaux, Alexandre Gramfort, Vincent Michel, Bertrand Thirion, Olivier Grisel, Mathieu Blondel, Peter Prettenhofer, Ron Weiss, Vincent Dubourg, et al. 2011. Scikit-learn: Machine learning in python. *the Journal of machine Learning research*, 12:2825–2830.

James W. Pennebaker, Ryan L. Boyd, Kayla N Jordan, and Kate Blackburn. 2015. The development and psychometric properties of liwc2015. In *Psychometrics manual for text analysis program LIWC2015*.

Jeffrey Pennington, Richard Socher, and Christopher D Manning. 2014. GloVe: Global vectors for word representation. In *Proceedings of EMNLP*, pages 1532–1543.

Víctor M. Prieto, Sergio Matos, Manuel Alvarez, Fidel Cacheda, and José Luís Oliveira. 2014. Twitter: a good place to detect health conditions. *PloS one*, 9(1):e86191.

Kevin Roitero, VDMSM Cristian Bozzato, and G Serra. 2020. Twitter goes to the doctor: Detecting medical tweets using machine learning and bert. In *Proceedings of the International Workshop on Semantic Indexing and Information Retrieval for Health from heterogeneous content types and languages (SIIRH 2020)*.

Alex Wang, Amanpreet Singh, Julian Michael, Felix Hill, Omer Levy, and Samuel Bowman. 2019. Glue: A multitask benchmark and analysis platform for natural language understanding. In *7th International Conference on Learning Representations, ICLR 2019*.

Camilla Wasserman, Christina W Hoven, Danuta Wasserman, Vladimir Carli, Marco Sarchiapone, Susana Al-Halabi, Alan Apter, Judit Balazs, Julio Bobes, Doina Cosman, et al. 2012. Suicide prevention for youth-a mental health awareness program: lessons learned from the saving and empowering young lives in europe (seyle) intervention study. *BMC public health*, 12(1):776.

Thomas Wolf, Lysandre Debut, Victor Sanh, Julien Chaumond, Clement Delangue, Anthony Moi, Pierric Cistac, Tim Rault, R'emi Louf, Morgan Funtowicz, and Jamie Brew. 2019. Huggingface's transformers: State-of-the-art natural language processing. *ArXiv*, abs/1910.03771.

World Health Organization. 2016. *World health statistics 2016: monitoring health for the SDGs sustainable development goals*. World Health Organization.

Yonghui Wu, Mike Schuster, Zhifeng Chen, Quoc V Le, Mohammad Norouzi, Wolfgang Macherey, Maxim Krikun, Yuan Cao, Qin Gao, Klaus Macherey, et al. 2016. Google's neural machine translation system: Bridging the gap between human and machine translation. *arXiv preprint arXiv:1609.08144*.

Zheng Xie, Chaitanya Gadepalli, and Barry MG Cheetham. 2017. Reformulation and generalisation of the cohen and fleiss kappas. *LIFE: International Journal of Health and Life-Sciences*, 3(3).

Andrew Yates, Arman Cohan, and Nazli Goharian. 2017. Depression and self-harm risk assessment in online forums. In *Proceedings of the 2017 Conference on Empirical Methods in Natural Language Processing*, pages 2968–2978.

Antonia Zapf, Stefanie Castell, Lars Morawietz, and André Karch. 2016. Measuring inter-rater reliability for nominal data–which coefficients and confidence intervals are appropriate? *BMC medical research methodology*, 16(1):93.

Identifying Medication Abuse and Adverse Effects from Tweets: University of Michigan at #SMM4H 2020

V.G.Vinod Vydiswaran,[1,2†] **Deahan Yu,**[2] **Xinyan Zhao,**[2] **Ermioni Carr,**[2]
Jonathan Martindale,[2] **Jingcheng Xiao,**[3] **Noha Ghannam,**[2] **Matteo Althoen,**[2]
Alexis Castellanos,[4] **Neel Patel,**[5] **Daniel Vasquez**[2]

[1]Department of Learning Health Sciences, University of Michigan Medical School
[2]School of Information, University of Michigan; [3]College of Pharmacy, University of Michigan
[4]College of Engineering & Computer Science, University of Michigan, Dearborn
[5]Internal Medicine Department, Garnet Health Medical Center
[†] Corresponding author: `vgvinodv@umich.edu`

Abstract

The team from the University of Michigan participated in three tasks in the Social Media Mining for Health Applications (#SMM4H) 2020 shared tasks – on detecting mentions of adverse effects (Task 2), extracting and normalizing them (Task 3), and detecting mentions of medication abuse (Task 4). Our approaches relied on a combination of traditional machine learning and deep learning models. On Tasks 2 and 4, our submitted runs performed at or above the task average.

1 Introduction

The Social Media Mining for Health Applications (#SMM4H) Shared Task 2020 provided a unique hands-on experience for graduate students and researchers to apply concepts in text cleaning, natural language processing, and health informatics on standardized research tasks. The tasks are motivated by challenges in using social media data for health research, including informal, colloquial expressions and misspelling of clinical concepts, data sparsity, noise, and ambiguity (Klein et al., 2020).

This year, a team of students from the University of Michigan participated in three tasks – on classification of tweets that report adverse effects (Task 2 English), extraction and normalization of adverse effects (Task 3), and characterization of chatter related to prescription medication abuse (Task 4). We submitted three runs for each of these tasks. This paper describes our approach in developing and training the components for our participation, along with the results over validation and test data sets.

2 Task 2: Classifying tweets that report adverse effects

This binary classification task aimed at distinguishing tweets that report a medication adverse effect (AE) (labeled "positive"), from those that do not (labeled "negative"), taking into account subtle linguistic variations between adverse effects and "indication" or the reason for using the medication (Weissenbacher et al., 2019). The training data (n=25,672) was split into train (n=20,544; 80%) and validation (n=5,134; 20%) sets. 2,374 (9.2%) tweets in the training data were "positive". The test data had 4,759 tweets.

We participated in the previous iteration of this task in #SMM4H 2019, and proposed two support vector machine (SVM) models and a bidirectional long short-term memory (LSTM) model (Vydiswaran et al., 2019). We followed up on our participation during the post evaluation phase to further develop our approaches and proposed a neural network (NN) model called the collocated LSTM with attentive pooling and aggregated representation (CLAPA) (Zhao et al., 2019), that performed better than the three official submissions. Based on those insights, this year we proposed three NN models, viz. a multi-stage convolutional neural network model, the CLAPA model, and a BERT-augmented CLAPA model.

Data cleaning and preprocessing: We removed all duplicate tweets between the training and validation data sets, removed URLs and non-alphanumeric characters, and normalized the Twitter user mentions by replacing them with a token "username". We also applied the `emoji` library[1] in Python to substitute the emoticons with synonymous descriptive text. The tweets were tokenized and padded to normalize them to the same length.

---

[1]The `emoji` library. Last accessed 2020 Jul 7. https://pypi.org/project/emoji/

Proceedings of the 5th Social Media Mining for Health Applications (#SMM4H) Workshop & Shared Task, pages 90–94
Barcelona, Spain (Online), December 12, 2020.

Run 1 – Convolutional Neural Network (CNN):   The CNN model consisted of an input representation using Twitter Word2Vec embeddings (Godin, 2019) fed into a convolutional layer of 128 filters and 5 kernels, followed by a global max pooling layer to reduce input dimensionality, and a dropout layer to reduce overfitting. A sigmoid activation function is used to output the final label. The model parameters were tuned over the validation set, and used to train the final model on train and validation sets.

Run 2 – CLAPA: We retrained a model we proposed last year, called CLAPA – a collocated LSTM model with attentive pooling and aggregated representation (Zhao et al., 2019). This model takes a concatenation of word embbedings and collocated embedding into an LSTM layer, and passes the final hidden states to an attentive pooling layer. These pooled states are aggregated with medical collocation information and fed into a fully connected layer for final prediction label.

Based on the model described by (Zhao et al., 2019), medical concepts were constructed over the training set and expanded with a list of drugs from MedlinePlus[2]. A collocation graph was constructed with nodes as unique words in the dataset, and undirected edges from each medical concept to its 15 closest word neighbors. FastText pretrained word embedding (Joulin et al., 2017) was used as the input representation. The model consists of one layer of LSTM network with the hidden size as 300 and three multi-head attention layers.

Run 3 – BERT-augmented CLAPA: As our third run, we developed an extension to CLAPA with Bidirectional Encoder Representations from Transformers (BERT) (Devlin et al., 2019). We extended CLAPA to utilize BERT's logits because BERT encodes a powerful, yet functionally different, representation for a given input. Specifically, the logits generated by BERT focus on representations from the full context of a tweet while the logits generated by CLAPA focus heavily on external medical domain information. While both CLAPA and BERT are independently trained on the dataset, in this BERT-augmented CLAPA model, we configure BERT to pass its logits to the CLAPA to allow the CLAPA model to make prediction based on the concatenation of the two logits.

3   Task 3: Extracting and normalizing adverse effects mentioned in tweets

This task involves extracting the span of text containing an AE of a medication from tweets that report an AE, and then mapping the extracted AE to a standard concept ID in the Medical Dictionary for Regulatory Activities (MedDRA) vocabulary (Mozzicato, 2009). The training data includes tweets that report an AE (annotated as "positive") and those that do not (annotated as "negative"). For each "positive" tweet, the training data contains the span of text containing the AE, the character offsets of that span of text, and the MedDRA ID of the AE. There were 2,376 tweets in the training set, of which 1,212 (51%) were "positive". The test set had 1,000 tweets.

Data cleaning and preprocessing: We removed all usernames, URLs, and non-alphanumeric characters in the tweets, and expanded all the contractions to normalize the text. To avoid downstream challenges of the preprocessing steps changing the character offsets of subsequent tokens, we created a character offset map for tokens in the raw tweet. Finally, we changed the tweet text to lowercase, and tokenized it using NLTK's `twitter_tokenizer` package.

We formulated the adverse effect extraction task as a sequence classification problem. We prepared the tokenized tweet text for this task by tagging each token in the training set with one of B-AE, I-AE, or O tags, to indicate the beginning, inside, and outside an AE span, respectively. Regular expression-based matching was used to identify the annotated spans in the training set.

Model development: All three runs that we submitted for Task 3 were based on a standard bidirectional long short-term memory (Bi-LSTM) model, followed by a conditional random field (CRF) model. However, the three models differed on the word embedding used to represent the tweet tokens, in model hyperparameters, and in the MedDRA Concept Unique Identifier (CUI) lists for the normalization task.

---

[2] MedlinePlus [Internet]. Bethesda (MD): National Library of Medicine (US); [updated 2020 Jul 6]. Drugs, Herbs and Supplements; [updated 2015 Apr 28; cited 2020 Jun 29]. Available from: https://medlineplus.gov/druginformation.html

Run 1 – Bi-LSTM CRF + MedDRA: For Run 1, we concatenated the representation from two word embbedings – viz. GloVe (Pennington et al., 2014) and EXT (Komninos and Manandhar, 2016) – and fed into a bi-LSTM model, followed by a CRF layer for this sequence classification task. The size of the hidden layer in the LSTM model was set as 75, the batch size was set as 64, and learning rate was 0.005.

To normalize the extracted text to a MedDRA term code, we applied QuickUMLS (Soldaini and Goharian, 2016) on the extracted text. QuickUMLS performs a fuzzy string match to find potential candidate CUI matches. When two or more CUIs were identified as candidates, the UMLS-MedDRA[3] list was used to choose the first candidate match as our normalized MedDRA CUI output.

Run 2 – Bi-LSTM CRF + MedDRA with different hyperparameters: The model architecture is similar to the one used in Run 1, except that we used only the EXT word embedding as the input representation. The size of the hidden layer in the LSTM model was set as 55, the batch was set as 32, and learning rate was 0.001. The approach used for normalization was identical to the one used in Run 1.

Run 3 – Bi-LSTM CRF with a modified version of MedDRA: For Run 3, the extraction model was identical to that used in Run 1, but the normalization method was different. Instead of the UMLS-MedDRA list, we used a modified version of the dictionary. This modified version was created by first identifying examples of the most commonly missed phrases in the training set, and then manually adding them based on their concept identifiers in the UMLS Metathesaurus (Bodenreider, 2004).

4   Task 4: Characterizing prescription medication abuse in tweets

This multi-class classification task involves distinguishing, among tweets that mention at least one prescription opioid, benzodiazepine, atypical anti-psychotic, central nervous system stimulant, or GABA analogue, tweets that report potential abuse/misuse (annotated as "A"), non-abuse/-misuse consumption (annotated as "C"), merely mention medications (annotated as "M"), or are unrelated (annotated as "U") (O'Connor et al., 2020). The training and test sets consist of 13,172 and 3,271 tweets, respectively.

The evaluation metric was set as the F1 score for the potential abuse/misuse class. So, we reformulate the task as a binary classification task of distinguishing potential abuse/misuse tweets (class "A") from non-abuse tweets (union of classes "C", "M", and "U").

Data cleaning and preprocessing: We removed all usernames, URLs, punctuation symbols, and special characters because they did not add substantial value to the classification of the tweet based on our initial analysis of the training set. We also removed all emojis as we observed that they were not as helpful, and often detrimental, to the training of our models, even when they were converted to descriptive tags such as ":smilingface:". We converted tweets to lower case and replaced all drug names, generic and brand name, with a token "drugname" so that models could focus on the context surrounding the mention of a drug, rather than the name of the drug itself.

Runs 1 & 3 – SVM with radial basis function (RBF) and linear kernels: Runs 1 and 3 were based on SVM models trained with an RBF kernel and a linear kernel, respectively. For both models, we set $C = 1.0$ and $\gamma = 0.5$. We also set the class weight parameter as "balanced" to adjust for the uneven distribution of classes in our binary abuse vs. non-abuse formulation. The model parameters were tuned over the validation set, and used to train the final model on train and validation sets. The primary difference between Runs 1 and 3 was that for Run 1, we explicitly balanced the abuse to non-abuse distribution in the train set by sub-sampling the non-abuse class during the training phase.

Run 2 – Sequential neural network: Run 2 was based on a sequential neural network composed of three dense layers utilizing relu activation. The input text was transformed into tf-idf weighted representation and fed into the first dense layer of size 16, followed by a dropout of 0.5. In the second dense layer, the number of training features was reduced to 4, followed by another dropout of 0.5. The output layer was configured to generate a binary output using sigmoid activation. We used a binary cross-entropy loss during optimization, using the Adam optimizer with a learning rate of 0.001 and a decay rate of $1e\text{-}6$.

---

[3]MDR (MedDRA) - Synopsis.   U.S. National Library of Medicine.   Last accessed 2020 Jul 8. https://www.nlm.nih.gov/research/umls/sourcereleasedocs/current/MDR/index.html

Task	Run description	Validation F1	Test set F1
Task 2	Run 1: CNN	0.510	0.35
	Run 2: CLAPA	0.603	0.44
	Run 3: BERT-augmented CLAPA	**0.629**	**0.51**
	Averaged best performance (n = 16)	*–*	*0.46*
Task 3: Extraction	Run 1: Bi-LSTM CRF + MedDRA	0.331	0.463
	Run 2: Bi-LSTM CRF + MedDRA w/ different hyperparameters	**0.350**	0.463
	Run 3: Bi-LSTM CRF + modified MedDRA	0.331	0.463
	Averaged best performance, Relaxed (n = 7)	*–*	*0.564*
Task 3: Normalization	Run 1: Bi-LSTM CRF + MedDRA	0.112	0.193
	Run 2: Bi-LSTM CRF + MedDRA w/ different hyperparameters	0.115	0.193
	Run 3: Bi-LSTM CRF + modified MedDRA	**0.154**	0.196
	Averaged best performance, Relaxed (n = 7)	*–*	*0.292*
Task 4	Run 1: SVM with RBF kernel	**0.46**	**0.49**
	Run 2: Sequential neural network	0.41	0.43
	Run 3: SVM with Linear kernel	0.45	0.47
	Averaged best performance (n = 4)	*–*	*0.49*

Table 1: Summary of the runs submitted by our team, and their F1 scores over validation and test sets. For Task 3, only the results corresponding to the "relaxed" token match are shown.

5 Results

The performance of our submitted runs for all three tasks on validation and test sets are summarized in Table 1. The table also includes the averaged F1 score of the best runs submitted by all participating teams in each task on the test set. In Task 2, the run using the BERT-augmented CLAPA model (Run 3) performed the best, and outperformed the averaged F1 score of best-performing runs submitted by Task 2 participants. We also note that the performance of our submitted runs on the test set were significantly lower than those on the validation set. In Task 3, all three runs we submitted performed similarly, which implies that the variations we tried in the word embedding used to represent the tweet tokens did not affect the overall performance. Run 3 performed slightly better on the normalization task, but was below the average F1 score for the task. Finally, in Task 4, the SVM model with RBF kernel (Run 1) performed the best among our submitted runs, and was at par with the average F1 score of the best-performing runs.

Task	Run description	Precision	Recall	F1
Task 2	Run 3: BERT-augmented CLAPA	**0.48**	0.54	**0.51**
	Averaged best performance (n=16)	*0.42*	*0.59*	*0.46*
Task 3: Extraction	Run 3: Bi-LSTM CRF + modified MedDRA	**0.806**	0.324	0.463
	Averaged best performance, Relaxed (n=7)	*0.607*	*0.557*	*0.564*
Task 3: Normalization	Run 3: Bi-LSTM CRF + modified MedDRA	**0.345**	0.137	0.196
	Averaged best performance, Relaxed (n=7)	*0.312*	*0.290*	*0.292*
Task 4	Run 1: SVM with RBF kernel	0.462	0.513	**0.49**
	Averaged best performance (n=4)	*–*	*–*	*0.49*

Table 2: Comparison of precision, recall, and F1 measures of the best runs submitted by our team on the test set, compared to the average of the best-performing runs from all participants in each task.

Table 2 shows additional details of our best-performing runs in terms of precision, recall, and F1 scores on the test set, and the corresponding measures averaged over the best-performing runs from all participants in each task. All of our best-performing runs, including those for Task 3, achieved a higher precision than the average precision of all submitted best-runs, but have lower recall. In future, we plan to further explore ways to improve the recall of our proposed approaches for all three tasks.

6 Conclusion

Our approach for participating in the #SMM4H 2020 Shared Tasks focused on building on our previous efforts in addressing related tasks. We participated in three tasks and experimented with a combination of traditional machine learning and deep learning models. Two of the submitted runs performed at or above the average task performance. Additional experiments are planned to further improve the recall, and thereby the overall performance, of our proposed models for these tasks.

References

Olivier Bodenreider. 2004. The Unified Medical Language System (UMLS): integrating biomedical terminology. *Nucleic Acids Research*, 32(suppl_1):D267-D270, January.

Jacob Devlin, Ming-Wei Chang, Kenton Lee, and Kristina Toutanova. 2019. BERT: Pre-training of deep bidirectional transformers for language understanding. In *Proceedings of the 2019 Conference of the North American Chapter of the Association for Computational Linguistics: Human Language Technologies, Volume 1 (Long and Short Papers)*, pages 4171–4186. Association for Computational Linguistics, June.

Fréderic Godin. 2019. *Improving and Interpreting Neural Networks for Word-Level Prediction Tasks in Natural Language Processing*. Ph.D. thesis, Ghent University, Belgium.

Armand Joulin, Edouard Grave, Piotr Bojanowski, and Tomas Mikolov. 2017. Bag of tricks for efficient text classification. In *Proceedings of the 15th Conference of the European Chapter of the Association for Computational Linguistics: Volume 2, Short Papers*, pages 427–431, Valencia, Spain, April. Association for Computational Linguistics.

Ari Z. Klein, Ilseyar Alimova, Ivan Flores, Arjun Magge, Zulfat Miftahutdinov, Anne-Lyse Minard, Karen O'Connor, Abeed Sarker, Elena Tutubalina, Davy Weissenbacher, and Graciela Gonzalez-Hernandez. 2020. Overview of the fifth Social Media Mining for Health Applications (#SMM4H) Shared Tasks at COLING 2020. In *Proceedings of the Fifth Social Media Mining for Health Applications (#SMM4H) Workshop & Shared Task*.

Alexandros Komninos and Suresh Manandhar. 2016. Dependency based embeddings for sentence classification tasks. In *Proceedings of the 2016 Conference of the North American Chapter of the Association for Computational Linguistics: Human Language Technologies*, pages 1490–1500, San Diego, California, June. Association for Computational Linguistics.

Patricia Mozzicato. 2009. MedDRA: An overview of the medical dictionary for regulatory activities. *Pharmaceutical Medicine*, 23(2):65–75.

Karen O'Connor, Abeed Sarker, Jeanmarie Perrone, and Graciela Gonzalez-Hernandez. 2020. Promoting reproducible research for characterizing nonmedical use of medications through data annotation: Description of a twitter corpus and guidelines. *Journal of Medical Internet Research*, 22(2):e15861.

Jeffrey Pennington, Richard Socher, and Christoper Manning. 2014. GloVe: Global vectors for word representation. In *Proceedings of the 2014 Conference on Empirical Methods in Natural Language Processing (EMNLP)*, pages 1532–1543, Doha, Qatar, October. Association for Computational Linguistics.

Luca Soldaini and Nazli Goharian. 2016. QuickUMLS: a fast, unsupervised approach for medical concept extraction. In *SIGIR Workshop on Medical Information Retrieval (MedIR)*, pages 1–4.

V.G.Vinod Vydiswaran, Grace Ganzel, Bryan Romas, Deahan Yu, Amy Austin, Neha Bhomia, Socheatha Chan, Stephanie Hall, Van Le, Aaron Miller, Olawunmi Oduyebo, Aulia Song, Radhika Sondhi, Danny Teng, Hao Tseng, Kim Vuong, and Stephanie Zimmerman. 2019. Towards text processing pipelines to identify adverse drug events-related tweets: University of Michigan @ SMM4H 2019 Task 1. In *Proceedings of the Fourth Social Media Mining for Health Applications (#SMM4H) Workshop & Shared Task*, pages 107–109, Florence, Italy, August. Association for Computational Linguistics.

Davy Weissenbacher, Abeed Sarker, Arjun Magge, Ashlynn Daughton, Karen O'Connor, Michael J. Paul, and Graciela Gonzalez-Hernandez. 2019. Overview of the fourth social media mining for health (SMM4H) shared tasks at ACL 2019. In *Proceedings of the Fourth Social Media Mining for Health Applications (#SMM4H) Workshop & Shared Task*, pages 21–30, Florence, Italy, August. Association for Computational Linguistics.

Xinyan Zhao, Deahan Yu, and V.G.Vinod Vydiswaran. 2019. Identifying adverse drug events mentions in tweets using attentive, collocated, and aggregated medical representation. In *Proceedings of the Fourth Social Media Mining for Health Applications (#SMM4H) Workshop & Shared Task*, pages 62–70, Florence, Italy, August. Association for Computational Linguistics.

How far can we go with just out-of-the-box BERT models?

Lucie M. Gattepaille

Uppsala Monitoring Centre, Uppsala, Sweden

`lucie.gattepaille@who-umc.org`[1]

Abstract

Social media have been seen as a promising data source for pharmacovigilance. However, methods for automatic extraction of Adverse Drug Reactions from social media platforms such as Twitter still need further development before they can be included reliably in routine pharmacovigilance practices. As the Bidirectional Encoder Representations from Transformer (BERT) models have shown great performance in many major NLP tasks recently, we decided to test its performance on the SMM4H Shared Tasks 1 to 3, by submitting results of pretrained and fine-tuned BERT models without more added knowledge than the one carried in the training datasets and additional datasets. Our three submissions all ended up above average over all teams' submissions: 0.766 F_1 for task 1 (15% above the average of 0.665), 0.47 F1 for task 2 (2% above the average of 0.46) and 0.380 F1 score for task 3 (30% above the average of 0.292). Used in many of the high-ranking submissions in the 2019 edition of the SMM4H Shared Task, BERT continues to be state-of-the-art in ADR extraction for Twitter data.

1 Introduction

Any medicinal product that has an effect, has the potential of causing harmful effects, in some individuals under some circumstances (Lindquist, 2008). Defined as 'a response to a drug that is noxious and unintended and occurs at doses normally used in man for the prophylaxis, diagnosis or therapy of disease, or for modification of physiological function' (WHO Meeting on International Drug Monitoring: the Role of National Centres, 1972), Adverse Drug Reactions (ADRs) are a public health concern and have been identified as the 5th leading cause of deaths within the European Union, with an estimated rate of 197,000 deaths per year and a cost of 79 billion euros within EU per year (Bouvy et al., 2015). Monitoring the safety of medicinal products over time, as they enter the market and get used in much more heterogenous populations than the clinical trials in which their safety was primarily assessed, is therefore essential for public health and is the goal of pharmacovigilance.

Traditionally, pharmacovigilance has relied on spontaneous reporting systems such as FAERS in the US and VigiBase, the WHO database if individual case safety reports, gathering reports of suspected ADRs from more than 130 national spontaneous reporting systems, including FAERS (Lindquist, 2003). Although these systems are particularly effective for detecting rare and serious ADRs, they suffer from limitations, notably under-reporting (Hazell and Shakir, 2006).

The appearance of social media platforms such as Twitter has provided pharmacovigilance with new data sources which could potentially complement spontaneous reports by the breadth of coverage of the populations (Sloane et al., 2015). Twitter alone was boasting 321 million active users as of February 2019 and thus could partially address the under-reporting problem spontaneous report systems suffer from. Nevertheless, it is yet unclear how social media could be integrated meaningfully into pharmacovigilance activities. The 2015-launched research consortium WEB-RADR (Ghosh and Lewis, 2015) has concluded that, under the current performance of methods for detection and extractions of ADRs, Twitter has limited value for pharmacovigilance (van Stekelenborg et al., 2019; Caster et al., 2018). Although, the methods they have employed are not today's state-of-the-art NLP methods (Gattepaille et

---

Proceedings of the 5th Social Media Mining for Health Applications (#SMM4H) Workshop & Shared Task, pages 95–100
Barcelona, Spain (Online), December 12, 2020.

al., 2020), it shows there is a need for developing performant algorithms for automatic detection, extraction and characterization of ADRs and their associated medicinal products.

To stimulate the development of such methods, the Social Media Mining for Health (SMM4H) has launched several Shared tasks over the years, with focus on specific aspects of the ADR extraction problem. This year, we participated in task 1 (binary classification for tweets containing a drug name or a dietary supplement), task 2 (binary classification for tweets containing an ADR mention) and task 3 (NER for ADR mentions and normalization to MedDRA® the Medical Dictionary for Regulatory Activities terminology is the international medical terminology developed under the auspices of the International Council for Harmonisation of Technical Requirements for Pharmaceuticals for Human Use (ICH)). As the Bidirectional Encoder Representations from Transformer (BERT) (Devlin et al., 2019) was proven to be powerful in the 2019 SMM4H Shared Task (Weissenbacher et al., 2019; Miftahutdinov et al., 2019), we decided to apply pre-trained BERT models everywhere, with some fine-tuning to the tasks data, to see how this simple approach, requiring no domain expertise, would lead us. In this document, we report on the results of this experiment.

2 Methods

2.1 Preprocessing

All tweets and text extracts across the different datasets and tasks were preprocessed in the same manner. We first lowercased the text, then converted URLs, user tags as well as numbers to the special URL, USER and NUMBER tags respectively. We separated but kept the hash from the hashtags, collapsed characters appearing at least 3 times in a row into one character. Finally, we separated non-alpha characters from all other characters by a white space and collapsed all multiple white spaces into one, so the final token list became white-space-separated. For use in BERT models, all the preprocessed tweets were then passed to the BERT-base-uncased tokenizer, and consequently padded or truncated to a single length which was task-dependent.

2.2 Task 1 submission

For this task, we added the SMM4H 2018 tweets annotated for presence or absence of drug names to the training set (Weissenbacher et al., 2018), resulting in a training set of 65,041 tweets. During preprocessing, we padded/truncated all tweets to a length of 88 'tokens' (tokens include words, punctuation and non-alpha characters, and the smaller word chunks created by the BERT tokenizer). We fine-tuned the pre-trained BERT-base-uncased model found in the *Transformers* Python library from HuggingFace (Wolf et al., 2019) on the extended training set for 4 epochs, using the binary cross-entropy loss, with a batch size of 12, 0.1 dropout, the Adam optimizer and a linearly decreasing learning rate starting at 2e-5. A tweet was classified as a drug tweet if its post-softmax score exceeded 0.99. Very little fine-tuning of the hyperparameters was done. We submitted only one system run on the test set.

2.3 Task 2 submission

For this task, we also used a simple pre-trained BERT-base-uncased model and fine-tuned it on Task 2 training data, as well as Task 3's training and validation data, where the tweet labels were computed based on the presence or absence of an ADR mention, leading to a training set of 23,350 tweets. All tweets were padded/truncated to a length of 88 BERT tokens. The parameters and settings used for the model were the following: batch size of 12, dropout of 0.05, binary cross-entropy loss, Adam optimizer with a linearly decreasing learning rate starting at 2e-5 and 4 training epochs. A tweet was classified as an ADR tweet if its post-softmax score exceeded 0.99. Very little fine-tuning of the hyperparameters was done. We submitted only one system run on the test set.

2.4 Task 3 submission

For this task, we also applied BERT, both for the NER part and the normalization part of the task. We used the BIO-labelling scheme. BERT representations for all tokens were obtained via a pre-trained BERT-base-uncased model and passed to a softmax layer to classify each token as B, I or O. We did not include additional training data and only used the tweets with at least one ADR mention. We padded/truncated the tweets to a length of 50 BERT tokens. We trained the model for 16 epochs, with a batch size of 12, the Adam optimizer with a linearly decreasing learning rate starting at 2e-5, 0.1 dropout, unweighted cross-entropy loss. All tokens predicted as I but preceded by a token predicted as O were converted to O. Separately, we trained a multi-class classifier based on another pre-trained BERT-base-uncased for the normalization part of the task. We combined the training ADR extracts with the CADEC (Karimi et al., 2015) and the SMM4H 2017 task 3 (Sarker and Gonzalez-Hernandez, 2017) datasets, leading to a total of 40,162 ADR-text and MedDRA PT code pairs for training, spanning over 674 unique PT codes. Text extracts were preprocessed and padded/truncated to a length of 40 BERT tokens. We topped the BERT output layer with a softmax layer on the 674 PT classes, and trained the entire model with cross-entropy loss, for 16 epochs, with an Adam optimizer and a fixed learning rate of 2e-5, a batch size of 12 and a dropout of 0.1. We submitted only one system run on the test set.

3   Results

3.1   Task 1

Our system performed above average on all metrics (F1, precision and recall), and was particularly performant in recalling the tweets with drug names or dietary supplements (Table 1). Although we did not apply a full grid search on the different hyperparameters of the model, we can still see some clear over-fitting to the validation set, especially regarding precision. As precision is the metric most influenced by prevalence of the positive class, a lower prevalence in the test set could partially explain a drop as well, but nothing in the task description indicated a lower prevalence of drug tweets in the test set. With only 35 positive examples in the validation set (thus presenting a high degree of imbalance), the chase for 'that extra positive example' can quickly have a strong effect on the generalization capabilities of a given architecture, leading us to believe that, that a particular 'improvement' to the model is nothing less than additional overfitting to the validation set.

	F1	Precision	Recall
Validation	0.82	0.7895	0.8571
Test	0.7665	0.7111	0.8312
Average Test	0.6646	0.7007	0.7039

Table 1: Comparison of performance for Task 1. The average test represents the performance metrics for the test dataset averaged across all Task 1 submissions.

3.2   Task 2

Our system performed above average in terms of F1-score but only slightly so (Table 2). It seemed clearly more geared towards precision, most likely owing to the very high threshold applied for ADR classification of a given tweet (0.99 on the post-softmax score). The large drop in performance between the validation and test performance came as a big surprise, considering that wedid very little hyperparameter tuning to the validation set, to avoid overfitting, that we kept the BERT approach rather similar across tasks and that the class imbalance was lower than in task 1 (at least on the validation set, with 0.2% drug tweets in task 1 against 9% ADR tweets in task 2). This seems to indicate systematic differences between the validation and test sets. A big difference in prevalence could explain in part the performance drop, although the biggest effect was observed on the recall metric which should, in theory, be more robust to prevalence effects. Without the ability to perform a proper analysis of the different types of errors made by the system on the test set, any explanation is pure speculation.

	F1	Precision	Recall
Validation	0.81	0.7412	0.8882
Test	0.47	0.58	0.40
Average Test	0.46	0.42	0.59

Table 2: Comparison of performance for Task 2. The average test represents the performance metrics for the test dataset averaged across all best submissions made by teams in Task 2.

3.3 Task 3

Our system performed above average on all relaxed metrics (F1, precision and recall), but only marginally so for the precision metric. The recall, on the other hand, exceeded the average by 51% (Table 3). The drop in performance between validation results and test results is less extensive than in task 2 but clearly more substantial than in task 1, maybe owing to the fact that overfitting happens both for the NER subtask and the normalization subtask. Our results on the strict metrics (when a true positive is obtained by matching the span of the ADR exactly and normalizing to the appropriate PT) were basically 0, revealing a likely error in indexing the start and end characters of the extractions.

	Metric type	F1	Precision	Recall
Validation	Relaxed	0.42	0.359	0.510
Test	Relaxed	0.380	0.335	0.439
Average Test	Relaxed	0.292	0.312	0.29

Table 3: Comparison of performance for Task 3. The average test represents the performance metrics for the test dataset averaged across all best submissions made by teams in Task 3.

Results of the NER subtask were also provided separately (Table 4). The simple BERT classifier on BIO labels was quite performant. The best performing system in the ADR NER task in the SMM4H 2019 (task 2) was using BioBERT (Lee et al., 2019) topped with a CRF and obtained and relaxed F1 score of 0.658 (0.554 precision and 0.81 recall) (Miftahutdinov et al., 2019), which is thus slightly lower than the performance of our system on this year's dataset. Although performance comparisons across different datasets should always be made with great caution, as one dataset may not be representative of the other, this shows that a simple pre-trained and fine-tuned BERT model can be powerful for this NER task, with relatively small amounts of data.

	Metric type	F1	Precision	Recall
Test	Relaxed	0.730	0.652	0.830
Average Test	Relaxed	0.564	0.607	0.557
Best system 2019	Relaxed	0.658	0.554	0.81

Table 4: Comparison of performance for the NER subtask of Task 3. The average test represents the performance metrics for the test dataset averaged across all best submissions made by teams in the NER subtask of Task 3.

4   Conclusion

"If all you have is a hammer, everything looks like a nail" (*Abraham Maslow*). As out-of-the-box pre-trained and fine-tuned BERT models have shown great performance in many kinds of NLP problems, we decided to pick up the BERT hammer and apply it to all tasks we registered for, to test its performance against the current efforts of the community for automated ADR extraction. Although we do not have the final rankings at the time of this writing, unfortunately, we see that our simple approach performed either slightly above (task 2) or largely above average (tasks 1 and 3). BERT has already been identified in the 2019 edition of the SMM4H Shared task as a contributor to good performance (Weissenbacher et al., 2019), as most high-ranking submissions in all tasks were using BERT or BioBERT in one way or another. We believe the picture is likely to be similar this year, although it will be interesting to see how the community integrated domain knowledge into their approaches and how such approaches fared against this submission, for which the domain knowledge is only included in the fine-tuning of the BERT algorithms.

Acknowledgements

MedDRA® trademark is registered by IFPMA on behalf of ICH.

References

Jacoline C. Bouvy, Marie L. De Bruin, and Marc A. Koopmanschap. 2015. Epidemiology of Adverse Drug Reactions in Europe: A Review of Recent Observational Studies. *Drug Safety*, 38(5):437–453, May.

Ola Caster, Juergen Dietrich, Marie-Laure Kürzinger, Magnus Lerch, Simon Maskell, G. Niklas Norén, Stéphanie Tcherny-Lessenot, Benoit Vroman, Antoni Wisniewski, and John van Stekelenborg. 2018. Assessment of the Utility of Social Media for Broad-Ranging Statistical Signal Detection in Pharmacovigilance: Results from the WEB-RADR Project. *Drug Safety*, 41(12):1355–1369, December.

Jacob Devlin, Ming-Wei Chang, Kenton Lee, and Kristina Toutanova. 2019. BERT: Pre-training of Deep Bidirectional Transformers for Language Understanding. *arXiv:1810.04805 [cs]*, May. arXiv: 1810.04805.

Lucie M. Gattepaille, Sara Hedfors Vidlin, Tomas Bergvall, Carrie E. Pierce, and Johan Ellenius. 2020. Prospective Evaluation of Adverse Event Recognition Systems in Twitter: Results from the Web-RADR Project. *Drug Safety*, May.

Rajesh Ghosh and David Lewis. 2015. Aims and approaches of Web-RADR: a consortium ensuring reliable ADR reporting via mobile devices and new insights from social media. *Expert Opinion on Drug Safety*, 14(12):1845–1853, December.

Lorna Hazell and Saad A W Shakir. 2006. Under-Reporting of Adverse Drug Reactions: A Systematic Review. *Drug Safety*, 29(5):385–396.

Sarvnaz Karimi, Alejandro Metke-Jimenez, Madonna Kemp, and Chen Wang. 2015. Cadec: A corpus of adverse drug event annotations. *Journal of biomedical informatics*, 55:73–81.

Jinhyuk Lee, Wonjin Yoon, Sungdong Kim, Donghyeon Kim, Sunkyu Kim, Chan Ho So, and Jaewoo Kang. 2019. BioBERT: a pre-trained biomedical language representation model for biomedical text mining. *Bioinformatics*:btz682, September.

Anna Marie Lindquist. 2003. *Seeing and observing in international pharmacovigilance: achievements and prospects in worldwide drug safety*. Uppsala Monitoring Centre, Uppsala.

Marie Lindquist. 2008. VigiBase, the WHO Global ICSR Database System: Basic Facts: *Drug Information Journal*, September.

Zulfat Miftahutdinov, Ilseyar Alimova, and Elena Tutubalina. 2019. KFU NLP team at SMM4H 2019 tasks: Want to extract adverse drugs reactions from tweets? BERT to the rescue. In *Proceedings of the Fourth Social Media Mining for Health Applications (# SMM4H) Workshop & Shared Task*, pages 52–57.

Abeed Sarker and Graciela Gonzalez-Hernandez. 2017. Overview of the Second Social Media Mining for Health (SMM4H) shared tasks at AMIA 2017. *Training*, 1(10,822):1239.

Richard Sloane, Orod Osanlou, David Lewis, Danushka Bollegala, Simon Maskell, and Munir Pirmohamed. 2015. Social media and pharmacovigilance: A review of the opportunities and challenges: Social media and pharmacovigilance. *British Journal of Clinical Pharmacology*, 80(4):910–920, October.

John van Stekelenborg, Johan Ellenius, Simon Maskell, Tomas Bergvall, Ola Caster, Nabarun Dasgupta, Juergen Dietrich, Sara Gama, David Lewis, Victoria Newbould, Sabine Brosch, Carrie E. Pierce, Gregory Powell, Alicia Ptaszyńska-Neophytou, Antoni F. Z. Wiśniewski, Phil Tregunno, G. Niklas Norén, and Munir Pirmohamed. 2019. Recommendations for the Use of Social Media in Pharmacovigilance: Lessons from IMI WEB-RADR. *Drug Safety*, 42(12):1393–1407, December.

Davy Weissenbacher, Abeed Sarker, Arjun Magge, Ashlynn Daughton, Karen O'Connor, Michael Paul, and Graciela Gonzalez. 2019. Overview of the fourth social media mining for health (SMM4H) shared tasks at ACL 2019. In *Proceedings of the Fourth Social Media Mining for Health Applications (# SMM4H) Workshop & Shared Task*, pages 21–30.

Davy Weissenbacher, Abeed Sarker, Michael Paul, and Graciela Gonzalez. 2018. Overview of the third social media mining for health (SMM4H) shared tasks at EMNLP 2018. In *Proceedings of the 2018 EMNLP Workshop SMM4H: The 3rd Social Media Mining for Health Applications Workshop & Shared Task*, pages 13–16.

WHO Meeting on International Drug Monitoring: the Role of National Centres (1971: Geneva and World Health Organization). 1972. *International drug monitoring: the role of national centres , report of a WHO meeting [held in Geneva from 20 to 25 September 1971].*World Health Organization technical report series ; no. 498. World Health Organization.

Thomas Wolf, Lysandre Debut, Victor Sanh, Julien Chaumond, Clement Delangue, Anthony Moi, Pierric Cistac, Tim Rault, R'emi Louf, Morgan Funtowicz, and Jamie Brew. 2019. HuggingFace's Transformers: State-of-the-art Natural Language Processing. *ArXiv*, abs/1910.03771.

FBK@SMM4H2020: RoBERTa for detecting medications on Twitter

Silvia Casola
Università di Padova
Fondazione Bruno Kessler
scasola@fbk.eu

Alberto Lavelli
Fondazione Bruno Kessler
lavelli@fbk.eu

Abstract

This paper describes a classifier for tweets that mention medications or supplements, based on a pretrained transformer. We developed such a system for our participation in Subtask 1 of the Social Media Mining for Health Application workshop, which featured an extremely unbalanced dataset. The model showed promising results, with an F_1 of 0.8 (task mean: 0.66).

1 Introduction

Twitter is a valuable source of user-generated data, including health data, and might be used for monitoring drug abuse and adverse effects online. However, tweets that mention medications need to be flagged first. To do so, lists of drugs are not enough: this is due to common spelling mistakes on social networks and language ambiguity (e.g. Xanax might be a drug or a Belgian band). The Substask 1 of the 2020 Social Media Mining for Health workshop (#SMM4H) challenged participants to identify tweets containing drugs or dietary supplements' mentions. The corpus showed a natural, highly imbalanced distribution of labels, with extremely rare positive occurrences. In this paper, we describe our entries, based on a pretrained RoBERTa model (Liu et al., 2019).

2 Datasets

The main dataset (DS2020), for which participants generated predictions, is a corpus of tweets by 112 pregnant women, as described in Weissenbacher et al. (2019). Its label distribution is unbalanced (181+/69091-), with positive instances only constituting 0.2% of the training set. Given the scarcity of positive samples, we also employed another dataset (DS2018), previously used in the 2018 #SMM4H shared tasks (Weissenbacher et al., 2018). It consists of 9611 tweets from 7584 users, annotated for medications' and supplements' mentions; the dataset was artificially balanced (4975+/4647-). Additionally, we used a large corpus of about 5 million general-domain, unlabelled English tweets.

3 Method

We submitted three runs to the leaderboard. All models were based on a RoBERTa pretrained transformer. RoBERTa is an enhancement of BERT (Devlin et al., 2019), with a modified training process (objective and data) and hyperparameters.

The first run (*Base model*) is a simple pretrained RoBERTa base uncased classifier. We first fine-tuned the original model on the DS2018 dataset to classify tweets mentioning medications and supplements. The hyperparameters were selected to maximize F_1 for the DS2020 validation set. Following a transfer learning approach, we used the resulting weights to initialize the final model and trained it on the DS2020 data. The original text was not preprocessed; for both datasets, we only used the text of the tweets, discarding the date, time, and user. No list of medications or other prior knowledge was employed and no explicit training on the medical domain was performed.

In the second run (*MLM model*), we followed a similar pipeline, but pretraining RoBERTa on a large corpus of tweets as a first step. The training was performed with a Masked Language Model (MLM)

Proceedings of the 5th Social Media Mining for Health Applications (#SMM4H) Workshop & Shared Task, pages 101–103
Barcelona, Spain (Online), December 12, 2020.

Figure 1: Our training (left) and test (right) pipeline for the *Ensemble MLM* model.

objective, virtually continuing the original transformer's training, to adapt it to the Twitter language specificities. Masked Language Modelling consists of masking out a fraction of the input tokens and training the model to predict the original tokens based on context.

Finally, we observed that, when trained for a varying number of epochs, the *MLM models* exhibited comparable F_1 but made different errors on the validation set. For this reason, we chose an ensemble of five MLM checkpoints (*Ensamble MLM model*) as our third run; we chose all checkpoints to have good results on validation data, but varying precision/recall thresholds. At test time, majority voting was used.

Once the hyperparameters were fixed on the validation set, all final models were trained on both the train and validation data to predict the test set. Figure 1 presents the pipeline employed for the third run.

4 Experiments and Results

In this section, we report our experimental results. Given the skewness of the dataset, system performances were measured in terms of F_1 for the positive class (i.e. tweets containing medications or supplements). Table 1 reports our results on the validation and test sets. The three runs surpassed by 14 points the mean of the submitted systems ($F_1 = 0.66$) and by 2 points the previous state of the art on the dataset (Weissenbacher et al., 2019), which used a complex ensemble model and external knowledge.

For all models, we used a batch size of 32, and experimentally found a constant learning rate of 10^{-6} to perform best. Surprisingly, a long fine-tuning seemed beneficial: training was stopped at epoch 132 (Base) and 73 (MLM) after evaluating the models at each epoch for a maximum of 200.

Figure 2 shows the errors made by the Base model. Note that some tweets might be hard to classify according to the gold standard even for a human annotator.

	Validation			Test		
	F_1	Precision	Recall	F_1	Precision	Recall
Base (DS2020)	0.81	0.82	0.8	-	-	-
Base (DS2018+ DS2020)	0.91	0.91	0.91	0.8	-	-
MLM model	0.86	0.84	0.89	0.8	-	-
MLM ensemble	0.88	0.84	0.91	0.8	0.77	0.83

Table 1: Experimental results

False positives: "Cayden's already starting to open his eyes! The anesthesia might wear off sooner than we thought...", "OBGYN's hate it when you tell them that pulling out is your form of birth control. Like HATE IT.", "I have a Man Cold. Well, I'm pregnant and can't take ALL the good drugs so it is like I have a Man Cold."

False negatives: "No first thing in the morning and still have this migraine!!! It's been now 4 days and nothing is working! Including a blood patch", "So I'm just about to drink castor oil again because at this point I'm tired", "Yeah im allergic to codiene idunno if I spelled that right I learned that the hard way".

Figure 2: Model errors

5 Conclusion

We presented our entries to the #SMM4H shared task. The model obtained promising results albeit trained on tweets' text only. Pretraining on the DS2018 set first highly improves the model performance (+0.1 F_1), while the MLM pretrain seems detrimental. This might be due to the quality of the unannotated dataset, which we used as obtained by the Twitter API, without any filtering or preprocessing.

References

Jacob Devlin, Ming-Wei Chang, Kenton Lee, and Kristina Toutanova. 2019. BERT: Pre-training of deep bidirectional transformers for language understanding. In *Proceedings of the 2019 Conference of the North American Chapter of the Association for Computational Linguistics: Human Language Technologies, Volume 1 (Long and Short Papers)*, pages 4171–4186, Minneapolis, Minnesota, June. Association for Computational Linguistics.

Yinhan Liu, Myle Ott, Naman Goyal, Jingfei Du, Mandar Joshi, Danqi Chen, Omer Levy, Mike Lewis, Luke Zettlemoyer, and Veselin Stoyanov. 2019. RoBERTa: A robustly optimized BERT pretraining approach. *arXiv preprint arXiv:1907.11692*.

Davy Weissenbacher, Abeed Sarker, Michael J. Paul, and Graciela Gonzalez-Hernandez. 2018. Overview of the third social media mining for health (SMM4H) shared tasks at EMNLP 2018. In *Proceedings of the 2018 EMNLP Workshop SMM4H: The 3rd Social Media Mining for Health Applications Workshop & Shared Task*, pages 13–16, Brussels, Belgium, October. Association for Computational Linguistics.

Davy Weissenbacher, Abeed Sarker, Ari Klein, Karen O'Connor, Arjun Magge, and Graciela Gonzalez-Hernandez. 2019. Deep neural networks ensemble for detecting medication mentions in tweets. *Journal of the American Medical Informatics Association*, 26(12):1618–1626.

Autobots Ensemble: Identifying and Extracting Adverse Drug Reaction from Tweets using Transformer Based Pipelines

Sougata Saha*, Souvik Das*, Prashi Khurana*, Rohini K. Srihari
Department of Computer Science and Engineering
University at Buffalo, Amherst, NY 14260
{sougatas, souvikda, prashikh, rohini}@buffalo.edu

Abstract

This paper details a system designed for Social Media Mining for Health Applications (SMM4H) Shared Task 2020. We specifically describe the systems designed to solve task 2: Automatic classification of multilingual tweets that report adverse effects, and task 3: Automatic extraction and normalization of adverse effects in English tweets. Fine tuning RoBERTa large for classifying English tweets enables us to achieve a F1 score of 56%, which is an increase of +10% compared to the average F1 score for all the submissions. Using BERT based NER and question answering, we are able to achieve a F1 score of 57.6% for extracting adverse reaction mentions from tweets, which is an increase of +1.2% compared to the average F1 score for all the submissions.

1 Introduction

With the world adapting to the new normal, social media is proving to be a key resource for humans. With more people sharing their life experiences in social media platforms, pharmaceutical firms can benefit by leveraging the power of deep learning and natural language processing for digital pharmacovigilance. In this paper, we showcase our systems for task 2 & 3 of the Social Media Mining for Health Applications Shared Task 2020 (Klein et al., 2020). Inspired by the current research using transformer architectures, and the results that KFU NLP Team (Miftahutdinov et al., 2019) had achieved at SMM4H 2019 using BERT (Devlin et al., 2018), we experimented with a suite of different transformer architectures. Transformers (Vaswani et al., 2017) are solely based on attention mechanisms, which dispense recurrence and convolutions entirely, enabling parallel processing and state of the art models. Liu et al. (2019) in their research uncovered that BERT was significantly under trained, and proposed RoBERTa: A Robustly Optimized BERT Pretraining Approach. We fine tuned RoBERTa on the English training tweets to classify a tweet as containing adverse reaction mention or not. For the Russian and French tweets classification tasks, we fine tuned RuBERT (Kuratov and Arkhipov, 2019) and CamemBERT (Martin et al., 2019) respectively.

For extracting the adverse reaction mentions from a tweet, we devised an end to end pipeline by posing the task as a named entity recognition (NER) task as well as a question answering task. We fine tuned BERT, SciBERT (Beltagy et al., 2019) and BioBERT (Lee et al., 2019) and created an ensemble NER, and fine tuned BioBERT QA for the question answering module. Post adverse mention extraction, we normalised the mention to the MedDRA code using pre-trained fastText (Bojanowski et al., 2016) embeddings and cosine similarity. The paper is organized as follows. In section 2 we describe the problem statements. Section 3 describes the architectures and methods that were implemented for each of the tasks, and showcase our results in section 4. We discuss some of the challenging aspects of the problems in section 5, and finally conclude in section 6.

Proceedings of the 5th Social Media Mining for Health Applications (#SMM4H) Workshop & Shared Task, pages 104–109
Barcelona, Spain (Online), December 12, 2020.

Task	Training Set			Test Set
	% Positive	% Negative	Total examples	
Task 2-English	9.25%	90.75%	25,672	5,000
Task 2-Russian	8.75%	91.25%	7,612	1,903
Task 2-French	1.61%	98.39%	2,426	607
Task 3-Resolution(NER + Norm)	51.01%	48.61%	2,376	1,000

Table 1: Data distribution for each task.

2 Task and Data Description

2.1 Task 2: Automatic classification of multilingual tweets that report adverse effects

This task involved developing a system which is capable of distinguishing tweets that report an adverse reaction to medication from tweets that do not. This task was subdivided into 3 tasks by language of tweet: English, Russian and French. Table 1 shows the distribution of training and testing data samples, and the split of positive and negative examples in the training data set.

2.2 Task 3: Automatic extraction and normalization of adverse effects in English tweets

This task consisted of two parts. The first being extraction of the specific adverse reaction of a drug from English tweets. The second being mapping the extracted adverse reaction to a standard concept ID in the MedDRA vocabulary. Table 1 shows the distribution of training and testing data samples, and the split of positive and negative examples in the training data set.

3 Methods

3.1 Task 2-Automatic classification of multilingual tweets that report adverse effects: English

Twitter data is almost always noisy, hence we cleansed and pre-processed the tweets before training the classifier. Using Ekphrasis (Baziotis et al., 2017), regex and NLTK, we converted tweets to lowercase, normalized elongated characters, repeated characters and hashtags, unpacked contractions, removed URL, mentions, smileys and emojis, and removed special tweet tokens like 'rt'.

We experimented with different transformer models like BERT base uncased, SciBERT with scivocab, BioBERT base v1.1 and RoBERTa large, and achieved best validation results with RoBERTa large. We fine tuned the RoBERTa large model using the pre-processed English tweets. We sum pooled the last 6 layers of RoBERTa, and performed classification by passing the pooled representation through a linear layer. We trained the model for 6 epochs with a learning rate of 2e-5. Table 2 demonstrates the model performance on the validation set, and Table 3 demonstrates the model performance on the test set.

3.2 Task 2-Automatic classification of multilingual tweets that report adverse effects: Russian and French

For the Russian and French classification, we pre-processed the tweets by removing special tweet tokens like 'rt', URL, mentions, smileys and emojis. We experimented with different transformer models like multilingual BERT, RuBERT, CamemBERT and FlauBERT (Le et al., 2019), and got best validation results using RuBERT for Russian tweets, and CamemBERT for French tweets. For both the models we had sum pooled the last 4 layers of the transformer, and trained for 6 epochs with a learning rate of 2e-5. For the French tweets we achieved a validation F1 of 0.22, but unfortunately could not classify any tweets correctly in the test data set. Table 2 demonstrates the performance of the Russian tweet classifier on the validation set, and Table 3 demonstrates the model's performance on the test set.

3.3 Task 3-Automatic extraction and normalization of adverse effects in English tweets: Extraction

We devised a three step extraction pipeline for this task, which is illustrated in Figure 1b. We detail the three steps below:

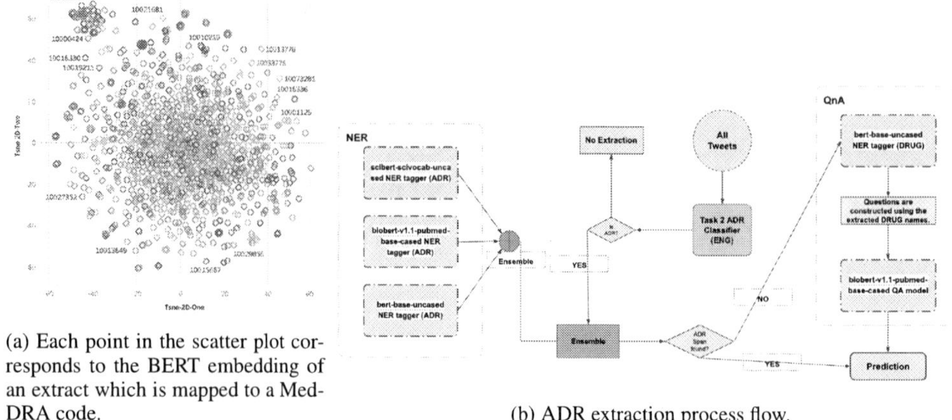

(a) Each point in the scatter plot corresponds to the BERT embedding of an extract which is mapped to a MedDRA code.

(b) ADR extraction process flow.

Figure 1: Pipeline for extracting adverse mentions from tweets and visualizing the adverse mentions extracts in 2-D using T-SNE

- **Classifying tweets as containing ADR mentions:** We cleanse the tweets using the same preprocessing pipeline as mentioned in section 3.1, and use the RoBERTa classifier trained from section 3.1 to classify tweets as containing adverse reaction mentions.

- **Transformer based ensemble named entity recognition (NER) tagger:** We fine tuned SciBERT, BioBERT and BERT base to create an ensemble of NER taggers for extracting the ADR mention extract. Each tweet was tagged using the 'BIO' scheme, where 'B' denoted the start token of an extract, 'O' denoted tokens outside the extract, and 'I' denoted tokens inside the extract. We fine tuned each of the models in the ensemble for 5 epochs with a learning rate of 3e-5. Only the tweets that are classified as containing adverse drug reaction mentions from the previous stage are passed through the ensemble NER tagger to extract the adverse mentions.

- **Transformer based question answering system:** The tweets that were classified as containing adverse mentions, but did not yield any extracts from the previous ADR mention extraction NER stage were passed through the question answering stage for adverse reaction extraction. This stage is sub divided into the following two steps:

  - **NER tagger for drug detection**: We trained a BERT NER tagger for detecting drug names in a tweet. We tagged each tweet using the standard 'BIO' scheme to distinguish tokens containing drug names from other tokens, and trained the classifier for 5 epochs with a learning rate of 3e-5. We passed tweets through this tagger to extract the drug name and passed the drug name and tweet to the below step.

  - **BioBERT question answering**: Given the tweet and the drug name as context, we fine tuned BioBERT question answering on our training dataset to extract the adverse reaction to the drug. For example, after identifying the drug name (for example Tylenol) through the drug NER tagger, we constructed the question "What is the adverse effect of Tylenol?". Given the constructed question and the tweet as context, we fine tuned the BioBERT question answering model for 3 epochs with a learning rate of 5e-6, to extract the adverse reaction mention from the tweet.

3.4 Task 3-Automatic extraction and normalization of adverse effects in English tweets: Normalization

The essence of this task was to assign the most probable MedDRA code to the extracted adverse reaction mention from a tweet. The distribution of the number of training examples per MedDRA code followed a long tail distribution, which led to majority of the MedDRA codes having insufficient training examples.

We overcame this problem by enriching the training data set with additional CADEC (Karimi et al., 2015) and UMLS (Bodenreider, 2004) adverse reaction to MedDRA code mapping pairs. We ensured the number of examples are approximately 50 for each MedDRA code. We leveraged pre-trained fastText word embeddings to map the extracted adverse reactions to the most probable MedDRA code. We denoted each of the 475 MedDRA codes by a 300 dimensional vector, which was computed by mean pooling the fastText word embeddings of all the 50 adverse reaction extracts associated with the MedDRA code. For each adverse reaction mention extract, we mean pooled the 300 dimensional fastText embedding of the extract and the tweet in a heuristically determined proportion of 10:1 and represented it as a fixed 300 dimensional vector. Finally cosine similarity between the 300 dimensional extract vector and all the 300 dimensional vectors for MedDRA codes was performed to determine the closest MedDRA code for the extract. The extract normalization process can be formulated by the following formulas.

During training:

$$x_{extracts} = \left(x_{extract\_1}, x_{extract\_2}, ..., x_{extract\_n} \right)$$

$$x_{extract\_i} = [x_1, x_2, ..., x_{300}]^T$$

$$x_{MedDRA\_code\_i} = 1/n * \sum_{x_{extract\_i} \in x_{extracts}} x_{extract\_i} = [\hat{x_1}, \hat{x_2}, ..., \hat{x_{300}}]^T$$

$$\mathbf{X}_{MedDRA\_code} = \left[x_{MedDRA\_code\_1}, ..., x_{MedDRA\_code\_475} \right]$$

During validation/testing:

$$x_{extract\_embedding} = [p_1, p_2, ..., p_{300}]^T$$

$$x_{tweet\_embedding} = [q_1, q_2, ..., q_{300}]^T$$

$$x_{extract\_embedding\_contextual} = [(10p_1 + q_1)/2, ..., (10p_{300} + q_{300})/2]^T$$

$$closest\_MedDRA\_code = argmax \ \frac{x_{extract\_embedding\_contextual}^T \bullet \mathbf{X}_{MedDRA\_code}}{norm(x_{extract\_embedding\_contextual}) * norm(\mathbf{X}_{MedDRA\_code}, dim = 0)}$$

4   Results

Strict and relaxed F1 scores are used for evaluating the models. Under strict mode of evaluation, ADR spans are considered correct only if both start and end indices matches with the indices in the gold standard annotations. Under relaxed mode of evaluation, ADR spans are considered correct only if spans in predicted annotations overlapped with the gold standard annotations. In our system, this leads to significant differences between the two F1 scores. Our RoBERTa based English tweet classifier for task 2 outperforms most systems, and achieves a test F1 score of 0.56. With the multi staged adverse reaction extraction pipeline, we are able to achieve a relaxed F1 score of 0.576, which is above the average F1 of all the other submitted systems. Unfortunately, due to highly imbalanced training data, our French tweet classifier is not optimally trained, and fails to correctly identify any French tweets containing adverse reaction mentions.

Below we summarize the performance of our systems on all the different tasks. Table 2 summarizes our system's performance on the validation set. In table 3 we summarize our system's performance on the test set, and also show a comparison between our system's performance, and the average performance of all the systems in each of the tasks.

Task	Strict			Relaxed		
	F1	Precision	Recall	F1	Precision	Recall
Task 2-English	0.648	0.630	0.668	-	-	-
Task 2-Russian	0.423	0.411	0.436	-	-	-
Task 3-Resolution(NER + Norm)	0.138	0.194	0.107	0.265	0.373	0.205

Table 2: Results on validation set.

Task	Strict			Relaxed		
	F1	**Precision**	**Recall**	**F1**	**Precision**	**Recall**
Task 2-English	**0.56**	**0.50**	**0.63**	-	-	-
Task 2-English(Avg others)	0.46	0.42	0.59	-	-	-
Task 2-Russian	0.36	0.3350	0.3976	-	-	-
Task 2-Russian(Avg others)	0.427	0.362	0.583	-	-	-
Task 3-Detection(NER)	0.291	0.317	0.269	**0.576**	**0.614**	0.543
Task 3-Detection(NER)(Avg others)	-	-	-	0.564	0.607	0.557
Task 3-Resolution(NER + Norm)	0.157	0.171	0.145	0.216	0.235	0.200
Task 3-Resolution(NER + Norm)(Avg others)	-	-	-	0.292	0.312	0.29

Table 3: Results on test set, and comparison against arithmetic mean of best submissions made by other teams.

5 Discussion

As represented in Table 1, although most of the classification tasks had imbalanced training data the transformer models performed well, as they are more resilient to imbalanced classes compared to traditional machine learning models. For the French tweet classifier, the models learning capabilities were seriously hampered by the highly imbalanced data set, and re-sampling techniques did not help.

Adverse drug reaction extract normalization was a particularly challenging task. We experimented with several hierarchical recurrent neural network based architectures, transformer architectures and fixed embedding based similarity architectures. We finalised on using embedding based similarity architecture as the other architectures did not boost the score much. In order to understand more about the problem, we looked into the data closely and uncovered the following patterns.

5.1 Extracts with similar meaning mapped to distinct MedDRA codes

There are extracts which are very similar in meaning, yet mapped to different MedDRA codes. For example, the extracts 'addiction' and 'addictive' are very similar in meaning, but are mapped to MedDRA codes 10001125 and 10012336 respectively. To make our normalization algorithm resilient to such differences, we included the embedding of the tweet as context, along with the embedding of the extract while mapping the extract to the MedDRA code. As discussed in section 3.4, we heuristically determined a weight of 10:1 for pooling the extract and tweet embedding to generate more contextual vector representation of the extract. Figure 1a illustrates the problem of overlapping extract embeddings in 2-D using T-SNE. We can see the extracts forming clusters, which makes the MedDRA code mapping problem hard.

5.2 Potential training data set issues

Consider the tweets 'addicted to nicotine badly' and '... dante addicted to that nicotine'. In both the tweets 'nicotine' is tagged as the drug, and 'addicted' is the adverse reaction extract. Intuitively, both the tweets should map to the same MedDRA code. However in the training data, the first tweet maps to MedDRA code 10012336, which stands for 'dependence addictive', and the second tweet maps to MedDRA code 10001125, which stands for 'addiction'. These examples hamper the learning capabilities of the models.

6 Conclusion

In this work, we have experimented with different transformer models for classifying ADR tweets as well as extracting ADR terms. We have leveraged several transformer based pre-trained models like RoBERTa, BioBERT etc. Also, we have devised a multi step pipeline for extracting the ADR terms from a tweet.

We can immediately think of two future improvements, firstly for the non English tasks in Task 2 we can develop a translation model to translate the tweets into English before classifying using the Task 2 English tweet classifier. Secondly, bettering the NER and MedDRA mapping, we want to incorporate a model that will be jointly trained to perform multiple tasks. For example, given a text, the model should be able to extract the ADR extracts as well as classify the tweet as ADR or non-ADR, as well as map it to the correct MedDRA code. Also, we will add a relationship extraction task, where we will identify the relation between the drug and ADR. We hypothesize that such a model should outperform a standard model as it will incorporate features and information sharing across tasks. For example, the NER would make a less false positive classification for non-ADR tweets.

References

Christos Baziotis, Nikos Pelekis, and Christos Doulkeridis. 2017. Datastories at semeval-2017 task 4: Deep lstm with attention for message-level and topic-based sentiment analysis. In *Proceedings of the 11th International Workshop on Semantic Evaluation (SemEval-2017)*, pages 747–754, Vancouver, Canada, August. Association for Computational Linguistics.

Iz Beltagy, Kyle Lo, and Arman Cohan. 2019. Scibert: A pretrained language model for scientific text.

Olivier Bodenreider. 2004. The unified medical language system (umls): integrating biomedical terminology. *Nucleic acids research*, 32(Database issue):D267–D270, Jan. 14681409[pmid].

Piotr Bojanowski, Edouard Grave, Armand Joulin, and Tomas Mikolov. 2016. Enriching word vectors with subword information.

Jacob Devlin, Ming-Wei Chang, Kenton Lee, and Kristina Toutanova. 2018. Bert: Pre-training of deep bidirectional transformers for language understanding.

Sarvnaz Karimi, Alejandro Metke-Jimenez, Madonna Kemp, and Chen Wang. 2015. Cadec: A corpus of adverse drug event annotations. *Journal of biomedical informatics*, 55:73–81.

Ari Z. Klein, Ilseyar Alimova, Ivan Flores, Arjun Magge, Zulfat Miftahutdinov, Anne-Lyse Minard, Karen O'Connor, Abeed Sarker, Elena Tutubalina, Davy Weissenbacher, and Graciela Gonzalez-Hernandez. 2020. Overview of the fifth social media mining for health applications (smm4h) shared tasks at coling 2020. *Proceedings of the Fifth Social Media Mining for Health Applications (SMM4H) Workshop & Shared Task*.

Yuri Kuratov and Mikhail Arkhipov. 2019. Adaptation of deep bidirectional multilingual transformers for russian language.

Hang Le, Loïc Vial, Jibril Frej, Vincent Segonne, Maximin Coavoux, Benjamin Lecouteux, Alexandre Allauzen, Benoît Crabbé, Laurent Besacier, and Didier Schwab. 2019. Flaubert: Unsupervised language model pretraining for french.

Jinhyuk Lee, Wonjin Yoon, Sungdong Kim, Donghyeon Kim, Sunkyu Kim, Chan Ho So, and Jaewoo Kang. 2019. Biobert: a pre-trained biomedical language representation model for biomedical text mining. *Bioinformatics*, Sep.

Yinhan Liu, Myle Ott, Naman Goyal, Jingfei Du, Mandar Joshi, Danqi Chen, Omer Levy, Mike Lewis, Luke Zettlemoyer, and Veselin Stoyanov. 2019. Roberta: A robustly optimized BERT pretraining approach. *CoRR*, abs/1907.11692.

Louis Martin, Benjamin Muller, Pedro Javier Ortiz Suárez, Yoann Dupont, Laurent Romary, Éric Villemonte de la Clergerie, Djamé Seddah, and Benoît Sagot. 2019. Camembert: a tasty french language model.

Zulfat Miftahutdinov, Ilseyar Alimova, and Elena Tutubalina. 2019. KFU NLP team at SMM4H 2019 tasks: Want to extract adverse drugs reactions from tweets? BERT to the rescue. In *Proceedings of the Fourth Social Media Mining for Health Applications (#SMM4H) Workshop & Shared Task*, pages 52–57, Florence, Italy, August. Association for Computational Linguistics.

Ashish Vaswani, Noam Shazeer, Niki Parmar, Jakob Uszkoreit, Llion Jones, Aidan N. Gomez, Lukasz Kaiser, and Illia Polosukhin. 2017. Attention is all you need.

Transformer Models for Drug Adverse Effects Detection from Tweets

Pavel Blinov, Manvel Avetisian
Sberbank Artificial Intelligence Laboratory, Moscow, Russia
{Blinov.P.D, Avetisyan.M.S}@sberbank.ru

Abstract

In this paper we present the drug adverse effects detection system developed during our participation in the Social Media Mining for Health Applications Shared Task 2020. We experimented with transfer learning approach for English and Russian, BERT and RoBERTa architectures and several strategies for regression head composition. Our final submissions in both languages overcome average F_1 by several percents margin.

1 Introduction

In the world of pandemic threats it is important to pay a special attention to the process of drug discovery. For the pharmaceutical industry it was always important not only to develop a new drug, but also to detect its possible Adverse Effects (AEs) even to smallest group populations as soon as possible. Although early stages of a clinical trial reveal most of a drug AEs, constant monitoring and feedback collection at the last pharmacovigilance stage is mandatory as it allows to identify rare or slowly evolving AEs. With current development of social media networks and artificial intelligence methods it is tempting to mine this information from the publicly available user data such as Twitter messages. If successful one can get an additional source of valuable information.

The above stated problem got into the research focus of 2020 Social Media Mining for Health Applications (SMM4H) Workshop (Klein et al., 2020). The organizers divided this problem into three consequent sub-tasks: 1) medical tweets detection; 2) classification of tweets that report AEs and 3) extraction and normalization of AEs from text. Specifically the second task was to build a binary classification model able to distinguish tweets that report an AE of a medication from those that do not. We concentrate our efforts only on this task as it is the only one that offers multilingual version (English, French and Russian).

In the recent years, methods for Natural Language Processing (NLP) developed rapidly. Nowadays transformer-based neural architectures (Vaswani et al., 2017) dominate the field. Our goal for this study was to benchmark BERT (Devlin et al., 2019) and RoBERTa (Liu et al., 2019) architectures and several strategies for regression head composition on this practical problem from medical domain with real-world data.

2 Data

The organizers provided labeled data alone with the Train / Validation split. Table 1 summarize statistics about the data for our languages of interest. As one can see the subset of the data for Russian language is very limited which affects the results (see Section 4).

In our experiments we don't use the provided split, but join train and validation parts for each language and perform 5-fold Cross-Validation (CV) procedure (Bishop, 2006). This allows us to estimate standard deviations for the models.

Proceedings of the 5th Social Media Mining for Health Applications (#SMM4H) Workshop & Shared Task, pages 110–112
Barcelona, Spain (Online), December 12, 2020.

Language	Train	Validation	Test
English	20,544 (9.4)	5,134 (9.84)	4,759 (-)
Russian	6,090 (8.75)	1,522 (8.74)	1,903 (-)

Table 1: Data statistics (number of examples (% of positive)).

The overall tendency of recent deep learning models is to avoid intermediate tasks and preprocessing, so the only preprocessing steps we perform with the labeled data is to remove user names and drop duplicate tweet messages (about 3.5% for the English).

3 Our approach

As the train data were very limited we resort to a transfer learning approach. That is, we take an NLP model pre-trained on a large corpus of texts (a backbone model) and fine-tune it for a specific task at hand. In this study as such backbone models for English we used BERT-base and RoBERTa-base from Transformers library (Wolf et al., 2019). For the Russian BERT model we fine-tuned RuBERT (Kuratov and Arkhipov, 2019). There was no Russian RoBERTa model, so we trained one from scratch on the public data (Shavrina and Shapovalova, 2017) and made it available.[1] This model was trained with Adam optimizer (starting learning rate 3×10^{-6}) on 16 GPU (Tesla V100) and the batch size of 1024 (with gradient accumulation) for about 60K steps.

Originally BERT and RoBERTa architectures designed a special classification token (cls) to summarize the state of a whole input text sequence (into a single embedding) and allow to configure downstream tasks (thought classification or regression head). It is known that in NLP-related tasks max-pooling operation over the embeddings associated with sentence keyword selection and the mean-pooling is a way of encoding the general meaning of the sentence. So we experimented with this pooling strategies over the last encoder states. Additionally the option with concatenation of embeddings (cls & max & mean) were explored. The resulting hidden state followed by additional fully connected layer of the same size and ends with regression output. Based on CV metric maximization strategy the threshold were selected to make final binarization of predictions.

4 Results and Conclusions

The primary evaluation metric for the task was F_1-measure toward the positive class as a trade-off between Precision (P) and Recall (R) (Manning et al., 2008). The 5-fold CV results of our experiments are shown in Table 2 (best regression heads for architecture and language are in **bold**).

		Regression head			
Language	Architecture	cls	max	mean	cls & max & mean
En	BERT	63.89 ± 1.3	$\mathbf{64.35 \pm 1.25}$	63.72 ± 1.33	64.02 ± 1.26
	RoBERTa	66.57 ± 0.19	66.63 ± 0.61	66.32 ± 1.2	$\mathbf{67.45 \pm 0.79}$
Ru	BERT	47.12 ± 3.94	47.87 ± 3.94	$\mathbf{48.64 \pm 3.82}$	48.1 ± 3.57
	RoBERTa	42.31 ± 3.98	43.01 ± 4.41	42.5 ± 4.74	$\mathbf{44.80 \pm 4.18}$

Table 2: 5-fold CV results for language, architecture and specific regression head ($F_1 \pm F_{\text{std}}$, %).

First of all, Table 2 shows that the English model performs much better with respect to the Russian one, probably because of the training data size (large standard deviation is in favor of this hypothesis). Second, the right choice of a regression head do help to get better performance.

For the English final submission we selected RoBERTa model with (cls & max & mean) regression head. Based on the test data it showed $F_1 = 57\%$ ($P = 50\%, R = 66\%$). That is better (by 11%) than average performing system $F_1 = 46\%$ ($P = 42\%, R = 59\%$).

---

[1]https://huggingface.co/blinoff/roberta-base-russian-v0

For the Russian submission we selected BERT model with mean regression head. In this case our system metrics is perfectly matched between CV and test, $F_1 = 48\%$ ($P = 36.1\%, R = 70.5\%$). Which is also 5.3% improvement compared to average scores for this task $F_1 = 42.7\%$ ($P = 36.2\%, R = 58.3\%$).

The results allow us to conclude that the drug adverse effects detection in raw text still remains challenging problem and requires more elaborate methods to reach the acceptable quality.

References

Christopher M. Bishop. 2006. *Pattern Recognition and Machine Learning (Information Science and Statistics).* Springer-Verlag, Berlin, Heidelberg.

Jacob Devlin, Ming-Wei Chang, Kenton Lee, and Kristina Toutanova. 2019. Bert: Pre-training of deep bidirectional transformers for language understanding. *Computing Research Repository*, arXiv:1810.04805. version 2.

Ari Z. Klein, Ilseyar Alimova, Ivan Flores, Arjun Magge, Zulfat Miftahutdinov, Anne-Lyse Minard, Karen O'Connor, Abeed Sarker, Elena Tutubalina, Davy Weissenbacher, and Graciela Gonzalez-Hernandez. 2020. Overview of the fifth social media mining for health applications (smm4h) shared tasks at coling 2020. *Proceedings of the Fifth Social Media Mining for Health Applications (SMM4H) Workshop Shared Task.*

Yuri Kuratov and Mikhail Arkhipov. 2019. Adaptation of deep bidirectional multilingual transformers for russian language. *Computing Research Repository*, arXiv:1905.07213.

Yinhan Liu, Myle Ott, Naman Goyal, Jingfei Du, Mandar Joshi, Danqi Chen, Omer Levy, Mike Lewis, Luke Zettlemoyer, and Veselin Stoyanov. 2019. Roberta: A robustly optimized BERT pretraining approach. *Computing Research Repository*, arXiv:1907.11692.

Christopher D. Manning, Prabhakar Raghavan, and Hinrich Schütze. 2008. *Introduction to Information Retrieval.* Cambridge University Press, USA.

Tatiana Shavrina and Olga Shapovalova. 2017. To the methodology of corpus construction for machine learning: "taiga" syntax tree corpus and parser. In *Proceedings of the International Conference CORPORA2017*, pages 78–84, Saint-Petersbourg, Russia.

Ashish Vaswani, Noam Shazeer, Niki Parmar, Jakob Uszkoreit, Llion Jones, Aidan N Gomez, Ł ukasz Kaiser, and Illia Polosukhin. 2017. Attention is all you need. In I. Guyon, U. V. Luxburg, S. Bengio, H. Wallach, R. Fergus, S. Vishwanathan, and R. Garnett, editors, *Advances in Neural Information Processing Systems 30*, pages 5998–6008. Curran Associates, Inc.

Thomas Wolf, Lysandre Debut, Victor Sanh, Julien Chaumond, Clement Delangue, Anthony Moi, Pierric Cistac, Tim Rault, R'emi Louf, Morgan Funtowicz, and Jamie Brew. 2019. Huggingface's transformers: State-of-the-art natural language processing. *Computing Research Repository*, arXiv:1910.03771.

Adverse drug reaction detection in Twitter using RoBERTa and Rules

Sedigheh Khademi, Pari Delir Haghighi, Frada Burstein
Monash University
{Sedigh.khademi, Pari.Delir.Haghighi, Frada.Burstein}@monash.edu

Abstract

This paper describes the method we developed for the Task 2 English variation of the Social Media Mining for Health Applications (SMM4H) 2020 shared task. The task was to classify tweets containing adverse effects (AE) after medication intake. Our approach combined transfer learning using a RoBERTa Large Transformer model with a rule-based post-prediction correction to improve model precision. The model's F1-Score of 0.56 on the test dataset was 10% better than the mean of the F1-Score of the best submissions in the task.

1 Introduction

The enormous quantity of self-reporting about medication effects in social media provides an opportunity for additional surveillance of medication effects, supplementing traditional health-related reporting systems (Edo-Osagie et al., 2020). Task 2 of the SMM4H Shared Task challenge (Klein et al., 2020) was defined as a binary classification task to identify tweets reporting an adverse effect (AE) from taking a medication. The language differences between AEs and indications (the reasons for using a medication) in the data could be quite subtle and increased the difficulty of the task. Datasets were available in English, French and Russian, but we only worked with the English dataset.

Fine tuning of state-of-the-art language models for classification has proven to be highly effective when applied to small but varied text collections (Weissenbacher et al., 2019). We chose a Hugging Face RoBERTa Large Transformer model (Liu et al., 2019) based on our previous experience in identifying vaccine adverse reaction mentions in tweets. Informed by previous studies that have used lexicons to improve classification results (Asghar et al., 2017), we developed a post-prediction rule-based correction based on a lexicon of phrases that mostly appeared in what we judged were non-AE tweets in the test data. This improved the F1-Score by 2%.

2 Data description and preparation

The Shared Task English data was supplied as two tab-delimited text files, one with 20,544 tweets as training data, and another of 5,134 tweets as a validation dataset. Each tweet was accompanied with a tweet id, a user id, and a "positive" class label of 1 for an AE, and a "negative" class label of 0 for a non-AE. Eventually an unlabeled test dataset of 4,759 tweets was provided by the challenge organizers for submission to the task. Around 9.3% of the data was designated as the positive AE class in the labeled datasets.

A comparison of tweets indicates how difficult this challenge is, there sometimes seemed to be little difference between an AE and something like it. For instance, these consecutive entries with their classes, both describing a side effect of pain following rivaroxaban intake: *negative class*: "04.40 just taken flecainide and 2 paracetamol to dull pain side effect of rivaroxaban."; *positive class*: "20.05 day 16 rivaroxaban diary. intense back and knee pain most of day; taken more paracetamol. had to shop, walking painful".

There were several aspects of the data which had a bearing on data preparation and classifier training, and on our conclusions about possible future directions: (i) We identified 423 texts that were in both the training and validation datasets, and (ii) within these datasets there were also duplicates; (iii) the data was highly imbalanced.

Proceedings of the 5th Social Media Mining for Health Applications (#SMM4H) Workshop & Shared Task, pages 113–117
Barcelona, Spain (Online), December 12, 2020.

To fix records that existed in both datasets, we removed them from the training data. We then also removed remaining duplicates from within each dataset. After cleaning, there were 1,896 positive labels in the training data and 473 in the validation data. The training dataset then consisted of 20,110 tweets – the 1,896 positive labels with the rest being all the negative labels. The de-duplicated validation dataset consisted of 2,167 negative and 473 positive labels, a total of 4,807 tweets.

We addressed the class imbalance in the training dataset by randomly sorting the data then splitting the negative labels of the training data into 7 datasets and combining each with all the positive labels. Therefore, each training dataset comprised of 2,602 negative labels and 1,896 positive labels, a total of 4,498 tweets.

The tweets text was not altered, and we added no features. Data was exported as a tab-delimited format suitable for use by the Transformer model. The Hugging Face libraries we used provided routines to tokenize and vectorize the text to Transformer model requirements.

3   Model

We used Hugging Face's RoBERTa Large Pytorch Transformer model (Liu et al., 2019) for classification, specifically the class *transformers.RobertaForSequenceClassification*, which is described in the Hugging Face documentation as a "RoBERTa Model transformer with a sequence classification/regression head on top (a linear layer on top of the pooled output)". RoBERTa is an acronym for "Robustly Optimized BERT Pretraining Approach" and was developed by Facebook to improve on Google's original BERT - "Bidirectional Encoder Representations from Transformers" (Devlin et al. 2019). A very summary explanation is that the models are based on Transformer language models and utilize multi-headed encoder/decoder attention mechanisms, which dispense with recurrence and convolutions entirely (Vaswani et al., 2017). They use intentionally masked sections of text to learn to predict the most probable words in sentences. RoBERTa improves on BERT by removing its next-sentence pretraining objective; by using larger mini-batches and learning rates; and using an order of magnitude more data and for a longer time than BERT was trained on. RoBERTa Large was the largest model of the available RoBERTa models on the Hugging Face site, we chose it because our tests on similar Twitter data have shown improvement in F1-Scores compared to BERT.

4   Method

4.1   Best model vs an ensemble

Models were learned on the larger imbalanced dataset and on the seven smaller balanced datasets, with F1-Scores calculated on the validation dataset. Experimentation showed that no more than 10 epochs were required to find the best model, and often only 2 epochs were needed. The best scoring model was retained from each of the eight datasets. An additional two models were kept because of their superior score, and the poorest scoring model was discarded, resulting in nine models. One of the retained models was learned on the large imbalanced dataset, the rest were learned on the smaller datasets, and the best individual performing model was one of these – it became our "Top 1" model. We then collated the predictions over the validation dataset for all odd-numbered (i.e. 3, 5, 7 and 9) groupings, to find out which maximum voting combinations most consistently matched the class. An ensemble of predictions from seven models had the best F1-Score - this was due to a reduction of false positives, leading to a greater precision. However, a consequence of the ensemble approach was a reduction in recall, and the single best model trained on one of the balanced datasets had the highest recall. Therefore, we considered how we might retain a high recall while reducing false positives to improve the F1-Score, and to assist with evaluation of recall significance we included an F1-beta score using a beta of 1.3.

4.2   Rule-based correction

Our data analysis included an examination of frequently appearing words and phrases in the classes. In the *training* data we noted that some phrases either inevitably or almost always were used by the negative class. Although this pattern did not appear consistently in the validation data, it was found in the test data. Analysis of our model predictions on the test data showed these phrases were present in what looked like false positive predictions. The phrases included mentions of lawyers and lawsuits, news and research, political figures, addictions, natural therapies, and encouragements to consult medical experts.

Therefore, we created post-prediction rules which we applied to the test data to enforce the negative class for tweets that contained these words or phrases.

For example, what we judged as the positive class and that our models had predicted as positive did *not* tend to contain phrases enjoining consultation with health experts or procuring medications. That is, phrases such as "ask your", "consult", "with your", "your doctor", "get your", and "try our" were almost always in discussions or in advertisements but *not* in reporting of adverse effects. To evaluate the effect of applying these rules in the test data we first labelled the test data with our model's predictions, then adjusted the labels to what the rules considered correct. This indicated that a 2% improvement in F1-Score was possible. We submitted a model with the correction, and in fact it did benefit by 2% over the same model without the correction.

5   Generated texts

When analyzing the data, we found two users in the training data who between them had 251 tweets that consisted of what looked like generated text. The tweets contained non-sensical phrases intermingled with key medication-related words, e.g.: "scuba dive ventolin hfa two brett butlers. the former"; only two of these texts were labelled as AE. We assessed the functionality of the 251 generated texts by learning classifiers on seven balanced datasets *without* the generated tweets, retaining the best model from each dataset. Compared to the models trained on all the data we found a higher recall to precision ratio, but that F1-scores were uniformly worse by at least 0.01. In this case we found an ensemble of the top 5 models gave the best F1-Score, but at 0.661 on the validation data it was 0.024 worse than the 0.685 of the best ensemble trained on all the data.

Although it was not performant, we wanted to know how this worked with the test data and so made a submission using the ensemble with a manual correction. Its test F1-Score was 0.46, which although identical with the mean best test dataset F1-Score for the challenge, was 10% below our best score of 0.56 obtained when the generated text was retained.

It was apparent that the generated text had a correcting effect by removing false positives. It seemed possible that this was due to their containing text with potentially AE indicative words surrounded with words and language structure that was not indicative of AE. A 10% difference was significant and so it seems likely that these texts have been introduced by the challenge organizers as a corrective measure.

6   Results

Scoring and Ensembles: Table 1 illustrates the scores obtained on validation data. The first column is the best model learned on the imbalanced dataset. Under the heading "Ensembles" the "Average" column is the average of the scores obtained from the models used in the various ensembles, and the remaining columns are scores obtained from ensembles, starting with "Top 1" which is the single best model trained on a balanced dataset.

	Imbalanced	Ensembles						Without
	dataset	Average	Top 1	Top3	Top 5	Top 7	Top 9	generated
TP	330	328	**366**	339	337	337	334	**380**
TN	4,146	4,139	4,083	4,147	4,149	4,160	**4,165**	4,037
FP	188	195	251	187	185	174	**169**	297
FN	143	145	**107**	134	136	136	139	**93**
Precision	0.637	0.627	0.593	0.644	0.646	0.659	**0.664**	0.561
Recall	0.698	0.693	**0.774**	0.717	0.712	0.712	0.706	**0.803**
F1-Score	0.666	0.659	0.672	0.679	0.677	**0.685**	0.684	0.661
F1-Beta (1.3)	0.674	0.667	**0.695**	0.688	0.686	0.692	0.690	0.692

Table 1: Model scores on validation data

The table shows that the best model for recall (F1-Beta at beta 1.3) is the "Top 1" model, but that the highest F1-Score belongs to an ensemble of 7 models due to its balance of recall and precision. The final

"Without generated" column contains the scores of the best, 5-model ensemble when models were trained without the generated texts – it has a higher recall than any other model, but also the poorest precision.

Submitted Models: Table 2 shows the test scores obtained from the models we submitted. The first column contains the means of the scores of the best models submitted to the task. The second column has all the scores from our best submitted model, which was the Top 1 model described above, but with the rule-based post-prediction corrective step applied. Its F1-Score it 10% better than the mean best F1-Score of the task. The third column is from the same model but without the corrections – as we had estimated there was a 2% difference between the two. The third column is the best model trained without the generated text in the training data – its F1-Score matches that of the mean best score. We were only given precision and recall scores for our best submission.

	Mean of Task best	Top 1 model corrected	Top 1 model uncorrected	Without generated
Precision	0.42	0.56		
Recall	0.59	0.55		
F1-Score	0.46	**0.56**	0.54	0.46

Table 2: Test scores of submitted models

7 Analysis

The best F1-Score of 0.685 we achieved on validation data outperformed the best F1-Score of 0.665 on validation data in the equivalent 2019 task. The model we submitted was our "Top 1" model with a manual correction applied, its (uncorrected) validation F1-Score of 0.672 was also above last year's best validation score. Our best F1-Score of 0.56 on the test data exceeded the average 2020 year's best test F1-Score of 0.46 by 10%.

An increased score over last year was possible as the RoBERTa Large model is superior to the BERT model used by some of last year's entrants. We did not do any extra training or tuning specific to the task (e.g. tuning with texts containing drug names from the training set). We did not evaluate any difference that might have been made by the removal of the duplicated tweets, but it is possible that some of our result is due to that data cleaning.

Under-sampling the negative class by splitting the data into 7 reasonably balanced datasets improved model performance, but it came with the price of decreasing the training examples. Ameliorating the data splitting by creating an ensemble of the best models' predictions increased precision, but, as noted above, this was at the expense of recall. In the end, the model used for our submission was the one that was trained on the best randomly created individual split dataset, as it had the highest recall and we wanted to evaluate manual corrections with it. We submitted predictions from this model with and without the rule-based post-prediction corrective step applied. The corrected model gained our top F1-Score of 0.56, the uncorrected score was 0.54. The extra 2% gained by the corrective step was useful and suggests additional development of such corrective processes.

The misclassified tweets on the validation data are numerous. Sometimes the difference in the language between the positive and negative classes was so subtle that neither we as non-experts, nor indeed the models, could pick the class correctly. For instance, all the models predicted the positive label for this negative class: "i am on only 25mg lamical but it makes me anxious and insomniac so i skip sometimes and get depressed". We can understand that the model finds anxiety, insomnia and depression as side effects, and although we as human judges can see that depression is in the context of not taking the medication the other two symptoms still pertain. However, we think that this was labelled as a negative class because it is not reporting a presently experienced side effect - the tense indicates an ongoing general experience with the medication.

8 Future work

Having established a benchmark, in the future we should try to understand the data further and see what can be done to improve the data as well as the models' ability to learn the data idiosyncrasies, as there is a lot of potential improvement in these areas. We are inspired by our observation of the effect of inserting generated texts into the data, and would like to research the applicability of that approach for improving training data - for instance, by inserting examples of the use of tense around side effects to indicate whether a current AE is being experienced. Additionally, we intend to explore the improvements possible with steps such as retraining the underlying language model with AE and applicable drug mentions, including using word masking on these texts and additional examples like them.

References

Asghar, M. Z., Khan, A., Ahmad, S., Qasim, M., & Khan, I. A. (2017). Lexicon-enhanced sentiment analysis framework using rule-based classification scheme. *PloS One, 12*(2), e0171649.

Edo-Osagie, O., De La Iglesia, B., Lake, I., & Edeghere, O. (2020). A scoping review of the use of Twitter for public health research. *Computers in Biology and Medicine*, 103770.

Klein, A. Z., Alimova, I., Flores, I., Magge, A., Miftahutdinov, Z., Minard, A.-L., O'Connor, K., Sarker, A., Tutubalina, E., Weissenbacher, D., & Gonzalez-Hernandez, G. (2020). Overview of the Fifth Social Media Mining for Health (SMM4H) Shared Tasks at COLING 2020. *Proceedings of the Fifth Social Media Mining for Health Applications (#SMM4H) Workshop & Shared Task*. https://doi.org/10.18653/v1/w19-3203

Liu, Y., Ott, M., Goyal, N., Du, J., Joshi, M., Chen, D., Levy, O., Lewis, M., Zettlemoyer, L., & Stoyanov, V. (2019). *RoBERTa: A Robustly Optimized BERT Pretraining Approach. 1*. http://arxiv.org/abs/1907.11692

Weissenbacher, D., Sarker, A., Magge, A., Daughton, A., O'Connor, K., Paul, M., & Gonzalez, G. (2019). Overview of the fourth social media mining for health (SMM4H) shared tasks at ACL 2019. *Proceedings of the Fourth Social Media Mining for Health Applications (# SMM4H) Workshop & Shared Task*, 21–30.

SpeechTrans@SMM4H'20: Impact of preprocessing and n-grams on Automatic Classification of Tweets that Mention Medications

Mohamed Lichouri
Computational Linguistics Department
CRSTDLA / Algiers-Algeria
m.lichouri@crstdla.dz

Mourad Abbas
Computational Linguistics Department
CRSTDLA / Algiers-Algeria
m.abbas@crstdla.dz

Abstract

This paper describes our system developed for automatically classifying tweets that mention medications. We used the Decision Tree classifier for this task. We have shown that using some elementary preprocessing steps and TF-IDF n-grams led to acceptable classifier performance. Indeed, the F1-score recorded was 74.58% in the development phase and 63.70% in the test phase.

1 Introduction

The 2020 Social Media Mining for Health Applications Workshop (Klein et al., 2020) launched several natural language processing tasks using social media mining for health monitoring for automatic classification of tweets: that mention medications, multilingual tweets that report adverse effects, tweets reporting a birth defect pregnancy outcome, in addition to automatic extraction and normalization of adverse effects in English tweets, and automatic characterization of chatter related to prescription medication abuse in tweets.

We are interested in the automatic classification of tweets that mention medication. A binary classification system has been experimented to achieve the task's aim, which is the distinction of tweets reporting medications from those that do not.

2 Dataset

In this section, we describe the dataset of task 1 that proposes to find tweets mentioning medications. Then, we present the pre-processing steps that we applied to clean the raw texts extracted from Twitter. The publicly available dataset (Weissenbacher et al., 2019) contains for each tweet: (i) the user ID, (ii) the tweet ID, and (iii) the binary annotation indicating the presence or absence of medications information. The dataset contains 69,272 tweets manually tagged. We noted that 0.26% of the dataset (181 tweets) mentioning medications are tagged as "positive", and 99.97% (69,091 tweets) that don't mention medication information are tagged as "negative".

3 System architecture

In our system, we applied three pre-processing steps. Where the first is the Tweets Preprocessor [1] developed by the AUTH team as part of the PlasticTwist Crowdsourcing module [2]. We used this tool to remove: all URLs, all mentions, all hashtags, Twitter reserved words (e.g. 'RT', 'via'), punctuation, single-letter words, blank spaces, stop-words, profane words, numbers. Whereas for the second step, we used Spacy tool [3] to parse the documents, and filter numbers, punctuation, white space, URL, while keeping the hashtag text. We also removed special characters, single-syllable tokens, mentions. We also

---

[1] https://github.com/vasisouv/tweets-preprocessor
[2] https://crowdsourcing.plastictwist.com/
[3] https://spacy.io/

Proceedings of the 5th Social Media Mining for Health Applications (#SMM4H) Workshop & Shared Task, pages 118–120
Barcelona, Spain (Online), December 12, 2020.

handled apostrophe, contraction check, spell correction as well as a lemmatization process as follows[4]:
We have done a *contraction check* to check if there is any contracted form, and replace it with its original
form (”aren't” is replaced by ”are not”) followed by a *Lemmatization process* where we lemmatize each
token using Spacy method '.lemma_', except for Pronouns, where they are kept as they are since Spacy
lemmatizer transforms every pronoun to ”-PRON-”. Finally, we will run a *Spell correction* to deals with
repeated characters such as ”sooooo gooooood”. Finally, for the third pre-processing step, we have used
some regex rules to remove punctuation and emojis.

In this work, we used a Machine Learning approach. In order to prepare the corpus, we adopted
three pre-processing steps which can be used individually or all together. After many experiments with
multiple combinations of the aforementioned three pre-processing steps, we decided to keep two choices
that gave the best performance: the 1st choice (applying steps 1 and 2 sequentially) and the 2nd choice
(steps 1, 2, 3).

After cleaning the data, we applied a TF-IDF vectorizer, and n-grams with multiple values of n (ranging from 3 to 20). We performed three tokenization process: word, character, and character with boundary (considers the space as a character) (Lichouri et al., 2018; Abbas et al., 2019). The classification has
been achieved using the Decision Tree algorithm. We applied re-sampling using k-Fold Cross-Validation
(Pedregosa et al., 2011). We set the values of k to 5 and 10.

4   Experiments and Results

As mentioned in the previous section, our system consists of applying multiple combinations of cleaners
(preprocessing steps) and using n-grams and tokenizers. We selected the three best results found for
development and test phase and addressed in table 1. Furthermore, we reported, in the same table, the
average performance of all teams of the task for the test set.

Dataset	Run ID	Configuration	Precision	Recall	F1-score
Dev	Run 1	steps (1,2,3) + 3-grams + 10-fold CV	73.53	**71.43**	72.46
	Run 2	steps (1,2,3) + 5-grams + 10-fold CV	88.00	62.86	73.33
	Run 3	steps (1,2) + 15-grams	**91.67**	62.86	**74.58**
Test	Run 1	steps (1,2,3) + 3-grams + 10-fold CV	-	-	62%
	Run 2	steps (1,2,3) + 5-grams + 10-fold CV	-	-	58%
	Run 3	steps (1,2) + 15-grams	**74.14%**	55.84%	63.70%
	Teams Avg.	-	70.32%	69.48%	66.28%

Table 1: Performance of the system in terms of precision, recall, and F1-score (*Dev and Test dataset*).

The size of the n-grams has an impact on the system's performance. Changing the value of n=3 to n=5,
led to an improvement of Precision and F1-score by more than 15% and 1%, respectively, while Recall
dropped out by more than 8%, as shown in table 1. The impact of preprocessing is noticeable through
the run 3 (see table 1). In fact, using preprocessing steps (1 and 2) and n-grams with n=15 sequentially
has given the best performance in the development phase with a precision of 91.67% and an F1-score of
74.58%. In the test phase, the third run still gives the best performance in terms of F1 score (63.70%)
with a precision of 74.14% compared to runs 1 and 2 (62% and 58%) respectively (Table 1).

5   Conclusion

The approach adopted in this work relies first on a set of preprocessing steps applied to the tweets'
dataset supplied in this task, and second, on the TF-IDF classifier with n-grams features, in addition
to tokenization module. We have shown that adequate choices of preprocessing steps combination and
values of n (n-grams) led to performance improvement. Compared to the average task performance, our
system gives an F1 score of 63.70% with a precision which outperforms the average by nearly 4%.

---

[4]https://github.com/tthustla/twitter_sentiment_analysis_part1

References

Mourad Abbas, Mohamed Lichouri, and Abed Alhakim Freihat. 2019. St madar 2019 shared task: Arabic fine-grained dialect identification. In *Proceedings of the Fourth Arabic Natural Language Processing Workshop*, pages 269–273.

Ari Z. Klein, Ilseyar Alimova, Ivan Flores, Arjun Magge, Zulfat Miftahutdinov, Anne-Lyse Minard, Karen O'Connor, Abeed Sarker, Elena Tutubalina, Davy Weissenbacher, and Graciela Gonzalez-Hernandez. 2020. Overview of the fifth social media mining for health applications (#smm4h) shared tasks at coling 2020. In *Proceedings of the Fifth Social Media Mining for Health Applications (SMM4H) Workshop Shared Task*.

Mohamed Lichouri, Mourad Abbas, Abed Alhakim Freihat, and Dhiya El Hak Megtouf. 2018. Word-level vs sentence-level language identification: Application to algerian and arabic dialects. *Procedia Computer Science*, 142:246–253.

F. Pedregosa, G. Varoquaux, A. Gramfort, V. Michel, B. Thirion, O. Grisel, M. Blondel, P. Prettenhofer, R. Weiss, V. Dubourg, J. Vanderplas, A. Passos, D. Cournapeau, M. Brucher, M. Perrot, and E. Duchesnay. 2011. Scikit-learn: Machine learning in Python. *Journal of Machine Learning Research*, 12:2825–2830.

Davy Weissenbacher, Abeed Sarker, Ari Klein, Karen O'Connor, Arjun Magge, and Graciela Gonzalez-Hernandez. 2019. Deep neural networks ensemble for detecting medication mentions in tweets. *Journal of the American Medical Informatics Association*, 26(12):1618–1626.

Want to Identify, Extract and Normalize Adverse Drug Reactions in Tweets? Use RoBERTa

Katikapalli Subramanyam Kalyan
Department of Computer Applications
NIT Trichy, India
kalyan.ks@yahoo.com

Sivanesan Sangeetha
Department of Computer Applications
NIT Trichy, India
sangeetha@nitt.edu

Abstract

This paper presents our approach for task 2 and task 3 of Social Media Mining for Health (SMM4H) 2020 shared tasks. In task 2, we have to differentiate adverse drug reaction (ADR) tweets from nonADR tweets and is treated as binary classification. Task 3 involves extracting ADR mentions and then mapping them to MedDRA codes. Extracting ADR mentions is treated as sequence labeling and normalizing ADR mentions is treated as multi-class classification. Our system is based on pre-trained language model RoBERTa and it achieves a) F1-score of 58% in task 2 which is 12% more than the average score b) relaxed F1-score of 70.1% in ADR extraction of task 3 which is 13.7% more than the average score and relaxed F1-score of 35% in ADR extraction + normalization of task 3 which is 5.8% more than the average score. Overall, our models achieve promising results in both the tasks with significant improvements over average scores.

1 Introduction

Social media platforms in particular, twitter are used extensively by common public to share their experiences which also includes health-related information like adverse drug reaction (ADR) they experience while consuming drugs. Adverse Drug Reaction refers to unwanted harmful effect following the use of one or more drugs. The abundant health-related social media data can be utilized to enhance the quality of services in health-related applications (Kalyan and Sangeetha, 2020b). Our team participated in task 2 and task 3 of SMM4H 2020 shared task (Klein et al., 2020). Task 2 aims at identifying whether a tweet contains ADR mention or not. Task 3 aims at extracting ADR mentions and then normalizing them to MedDRA concepts.

2 Task 2 – Identification of ADR tweets

2.1 Problem Definition and Dataset

Task 2 aims at identifying whether a tweet contains ADR mention or not. An example of an ADR tweet is: *'thank god for vyvanse #addicted'*. Here *'addicted'* is the adverse drug reaction that happened because of consumption of the drug *'vyvanse'*. An example of a nonADR tweet is *'never take paxil #js'*. In this task, we learn a classification model which outputs the label 1 or 0 for a given tweet depending on whether it contains ADR mentions or not. In this dataset, the training set consists of 20544 tweets (18641 nonADR and 1903 ADR tweets), validation set consists of 5134 tweets (4660 nonADR and 474 ADR tweets) and the test set consists of 4759 tweets.

2.2 Methodology

Pre-processing

We apply the following pre-processing steps

Proceedings of the 5th Social Media Mining for Health Applications (#SMM4H) Workshop & Shared Task, pages 121–124
Barcelona, Spain (Online), December 12, 2020.

- Lowercase the text and remove consecutively repeating characters in the words (e.g., feeeeel → feel).

- Remove urls, @user mentions, retweet tag (rt), non-ASCII and punctuation characters.

- Expand English contractions (e.g., can't → cannot) and replace interjections with their meanings (e.g., ouch, oww → pain).

- Replace character smiley (e.g., :) → happy) and emoji (e.g., ☻ → grinning face) with their text descriptions.

Model Description

In recent times, the evolution of pretrained language models like BERT (Devlin et al., 2019), RoBERTa (Liu et al., 2019) changed the scenario in natural language processing from training downstream models from scratch to just fine-tuning pre-trained models. These models learn universal language representations from large training corpus and they can be used in downstream tasks by adding one or two layers which are specific to the task (Qiu et al., 2020). Our model is based on RoBERTa. We add task-specific sigmoid layer on the top of RoBERTa and then fine-tune the entire model using training dataset. We consider the final hidden state vector $t_{<s>} \in \mathbb{R}^h$ of the special token <s>as tweet representation. Here h represents hidden state vector size in RoBERTa-base and it is equal to 768. The vector $t_{<s>}$ is passed through sigmoid layer to get the prediction $\hat{y}$. Overall, the label $\hat{y}$ is computed as :

$$t_{<s>} = RoBERTa(tweet) \tag{1}$$

$$\hat{y} = Sigmoid(t_{<s>}W^T + b) \tag{2}$$

Here $W \in \mathbb{R}^{1 \times h}$ and $b \in \mathbb{R}$ are learnable parameters of sigmoid layer.

2.3 Experiments and Results

As ADR tweets are less in number compared to nonADR tweets in the training set, we augment training set with ADR tweets from SMM4H 2017 (Sarker et al., 2018) and SMM4H 2019 (Weissenbacher et al., 2019) ADR tweets classification datasets. Further, we include only randomly chosen 90% of nonADR tweets in the training set. By conducting random search over the range of hyperparameters values, we arrive at the following optimal set of hyperparameter values: batch size = 128, learning rate = 3e-5, dropout = 0.2 (applied on $t_{<s>}$ vector to reduce overfitting) and epochs = 10. We implement our model in PyTorch framework using transformers library from huggingface (Wolf et al., 2019).

Model	Precision	Recall	F1
RoBERTa (ours)	52.00	65.00	**58.00**
Average scores	42.00	59.00	46.00

Table 1: Task 2 - ADR Tweet classification results on test data

We report performance of our model and average scores in task 2 – ADR tweets classification in Table 1. Our model achieves an F1-score of 58% and it is 12% more than the average score.

3 Task 3 – Extract and Normalize ADR Mentions

3.1 Problem Definition and Dataset

This task involves ADR extraction followed by normalization. In the first part (ADR Extraction), to extract ADR mentions the model has to identify ADR tweets and then extract ADR mentions by identifying the spans in tweets. A tweet can have more than one ADR mention also and an ADR mention can be a sequence of words also. Example of a tweet with ADR mentions

@coolpharmgreg i don't care if they are toxic haha putting the cipro drops in is essentially equivalent to torture #oww

Here *'oww'*, *'toxic'* and *'equivalent to torture'* are the adverse drug reactions due to the consumption of the drug *'cipro'*. In the second part (ADR normalization), the extracted ADR mentions are mapped to the standard concepts in MedDRA vocabulary. In the above example, the ADR mentions *'oww'*, *'toxic'*, and *'equivalent to torture'* are mapped to the concepts *'pain (10033371)'*, *'drug toxicity (10013746)'* and *'feeling unwell (10016370)'* respectively.The dataset for this task consists of training set with 1862 tweets (1080 ADR tweets with 1464 ADR mentions and 782 nonADR tweets), validation set with 428 tweets (233 ADR tweets with 365 ADR mentions and 195 nonADR tweets) and test set with 976 tweets.

3.2 Methodology

ADR extraction is viewed as sequence labeling which is nothing but assigning a label to each of the tokens in the sequence. We follow BIO tagging: the tag 'B-ADR' represents tokens at the beginning of ADR mention, 'I-ADR' represents tokens inside ADR mention and 'O' represents nonADR tokens. We experiment with two models for this task. The first model is based on RoBERTa i.e., RoBERTaForTokenClassification. The second model is multi-task learning based RoBERTa. In this task, ADR tweet identification is the auxiliary task and ADR extraction is the main task. As these two tasks are similar in nature, by joint learning, the knowledge gained in auxiliary task improves the performance of the main task ADR extraction (Caruana, 1997; Crichton et al., 2017). Following the recent work in normalizing medical concepts (Kalyan and Sangeetha, 2020a; Subramanyam and Sivanesan, 2020), we treat concept normalization as multi-class classification and experiment with RoBERTa.

3.3 Experiments and Results

For ADR extraction, in case of RoBERTa based model, we set batch size = 64, epochs = 20 and learning rate = 0.00003. In case of multitask learning based RoBERTa model, $loss = \lambda \times L_{ADE} + (1-\lambda) \times L_{ADR}$. Here L_{ADE} represents loss related to ADR extraction and L_{ADR} is loss related to ADR detection. The value of λ is set to 0.8. Here, the model is trained for 30 epochs with learning rate = 3e-5 and batch size = 64. For ADR normalization, we use a learning rate of 3e-5 and batch size of 128. For both ADR extraction and normalization, we use AdamW optimizer (Loshchilov and Hutter, 2019).

Model	Evaluation	Type	Precision	Recall	F1
RoBERTa	NER	Relaxed	63.0	78.9	**70.1 (↑ 13.7)**
		Strict	41.1	54.2	46.8
	NER + Norm	Relaxed	30.4	39.8	34.5
		Strict	23.6	31.1	26.8
RoBERTa+MTL	NER	Relaxed	65.1	72.8	68.7
		Strict	45.2	52.5	48.6
	NER + Norm	Relaxed	32.6	37.7	**35.0 (↑ 5.8)**
		Strict	25.5	29.6	27.4
Average scores	NER	Relaxed	60.7	55.7	56.4
	NER + Norm	Relaxed	31.2	29.0	29.2

Table 2: Task 3 - ADR Extraction and Normalization results on test data. Here NER represents ADR Extraction and Norm represents ADR Normalization.

The performance of our models is reported in Table 2. From the table we observe that, a) RoBERTa based model achieved relaxed F1 score of 70.1% in ADR extraction which is 13.7% more than the relaxed average score b) Multi-task learning RoBERTa based model achieved relaxed F1 score of 35% in NER+Norm which is 5.8% more than the relaxed average score.

4 Conclusion

In this work, we explored the effectiveness of RoBERTa to identify, extract and normalize ADR mentions in tweets. In both task 2 and task 3, our proposed models achieved promising results with significant improvements over average scores.

References

Rich Caruana. 1997. Multitask learning. *Machine learning*, 28(1):41–75.

Gamal Crichton, Sampo Pyysalo, Billy Chiu, and Anna Korhonen. 2017. A neural network multi-task learning approach to biomedical named entity recognition. *BMC bioinformatics*, 18(1):368.

Jacob Devlin, Ming-Wei Chang, Kenton Lee, and Kristina Toutanova. 2019. Bert: Pre-training of deep bidirectional transformers for language understanding. In *Proceedings of the 2019 Conference of the North American Chapter of the Association for Computational Linguistics: Human Language Technologies, Volume 1 (Long and Short Papers)*, pages 4171–4186.

Katikapalli Subramanyam Kalyan and S Sangeetha. 2020a. Bertmcn: Mapping colloquial phrases to standard medical concepts using bert and highway network. Technical report, EasyChair.

Katikapalli Subramanyam Kalyan and S Sangeetha. 2020b. Secnlp: A survey of embeddings in clinical natural language processing. *Journal of biomedical informatics*, 101:103323.

Ari Z. Klein, Ilseyar Alimova, Ivan Flores, Arjun Magge, Zulfat Miftahutdinov, Anne-Lyse Minard, Karen O'Connor, Abeed Sarker, Elena Tutubalina, Davy Weissenbacher, and Graciela Gonzalez-Hernandez. 2020. Overview of the fifth social media mining for health applications (smm4h) shared tasks at coling 2020. In *Proceedings of the Fourth Social Media Mining for Health Applications (# SMM4H) Workshop & Shared Task*.

Yinhan Liu, Myle Ott, Naman Goyal, Jingfei Du, Mandar Joshi, Danqi Chen, Omer Levy, Mike Lewis, Luke Zettlemoyer, and Veselin Stoyanov. 2019. Roberta: A robustly optimized bert pretraining approach. *arXiv preprint arXiv:1907.11692*.

Ilya Loshchilov and Frank Hutter. 2019. Decoupled weight decay regularization. In *International Conference on Learning Representations*.

Xipeng Qiu, Tianxiang Sun, Yige Xu, Yunfan Shao, Ning Dai, and Xuanjing Huang. 2020. Pre-trained models for natural language processing: A survey. *arXiv preprint arXiv:2003.08271*.

Abeed Sarker, Maksim Belousov, Jasper Friedrichs, Kai Hakala, Svetlana Kiritchenko, Farrokh Mehryary, Sifei Han, Tung Tran, Anthony Rios, Ramakanth Kavuluru, et al. 2018. Data and systems for medication-related text classification and concept normalization from twitter: insights from the social media mining for health (smm4h)-2017 shared task. *Journal of the American Medical Informatics Association*, 25(10):1274–1283.

Kalyan Katikapalli Subramanyam and Sangeetha Sivanesan. 2020. Deep contextualized medical concept normalization in social media text. *Procedia Computer Science*, 171:1353 – 1362. Third International Conference on Computing and Network Communications (CoCoNet'19).

Davy Weissenbacher, Abeed Sarker, Arjun Magge, Ashlynn Daughton, Karen O'Connor, Michael Paul, and Graciela Gonzalez. 2019. Overview of the fourth social media mining for health (smm4h) shared tasks at acl 2019. In *Proceedings of the Fourth Social Media Mining for Health Applications (# SMM4H) Workshop & Shared Task*, pages 21–30.

Thomas Wolf, Lysandre Debut, Victor Sanh, Julien Chaumond, Clement Delangue, Anthony Moi, Pierric Cistac, Tim Rault, Rémi Louf, Morgan Funtowicz, et al. 2019. Huggingface's transformers: State-of-the-art natural language processing. *ArXiv*, pages arXiv–1910.

Detecting Tweets Reporting Birth Defect Pregnancy Outcome using Two-View CNN RNN based Architecture

Saichethan Miriyala Reddy
Indian Institute of Information Technology
Bhagalpur, Bihar, India
`miriyala.cse.1725@iiitbh.ac.in`

Abstract

This research work addresses a new multi-class classification task (fifth task) provided at the fifth Social Media Mining for Health Applications (SMM4H) workshop. This task involves distinguishing three classes of tweets that mention birth defects. We propose a novel two view based CNN-BiGRU based architectures for automatic tweet classification task. Experimental evaluation of our proposed architecture over the validation set gives encouraging result as it improves by approximately 7% over our single view model for the fifth task. Code of our proposed framework is made available on Github[1]

1 Introduction

Twitter is a micro-blogging system, which allows its users to publish tweets of up to 280 characters in length to tell others what they are doing, what they are thinking, or what is happening around them. Over the past few years twitter has become very popular, for every second, on average, around 6,000 tweets are tweeted on Twitter. Twitter is also considered as "It's what's happening". Social media can be considered as an enormous corpus since last decade many researcher started exploring different tasks such as sentiment analysis(Kouloumpis et al., 2011) (Wang et al., 2012), Named Entity Recognition(Baldwin et al., 2015) (Suman et al., 2020) and Disambiguation(Dredze et al., 2016) using this data.

More recently twitter data is used to identify and study a small cohort of Twitter users whose pregnancies with birth defect outcomes (Klein et al., 2019). Due to unique characteristics of social media there are few challenges associated with using social media data for health care research(Weissenbacher et al., 2019), including informal texts, colloquial expressions and misspellings of clinical concepts, noisy text, data sparsity, ambiguity, and multilingual posts.

2 Proposed Framework

Hashtags are one the most important aspects of Twitter. Hashtag lets users to apply dynamic, user-generated tagging that helps other users easily find posts on twitter with a specific theme or content. Users create and use hashtags by placing a hash symbol in front of a word or unspaced phrase in a tweet. The hashtag may contain symbols, digits and letters(capitalized & uncapitalized). Many tweets contain at least one hashtag, and often they provide context which motivated us to incorporate them in our framework.

2.1 Two View

We propose a novel two view CNN-RNN framework for this multi label classification task. The illustration of the Two view CNN-RNN framework is shown in Fig. 2. It contains two parts: The tweet part extracts semantic representations from clean text; the hashtag part which extracts semantic representation from hashtags. Concatenation of these semantic representations are fed to a dense layer and its output

---

[1] `https://github.com/Saichethan/SMM4H-2020`

Proceedings of the 5th Social Media Mining for Health Applications (#SMM4H) Workshop & Shared Task, pages 125–127
Barcelona, Spain (Online), December 12, 2020.

is used as an input to softmax classifier. We also used single view for comparison which has only tweet part as shown in Fig. 1, same hyperparameters are used for both models.

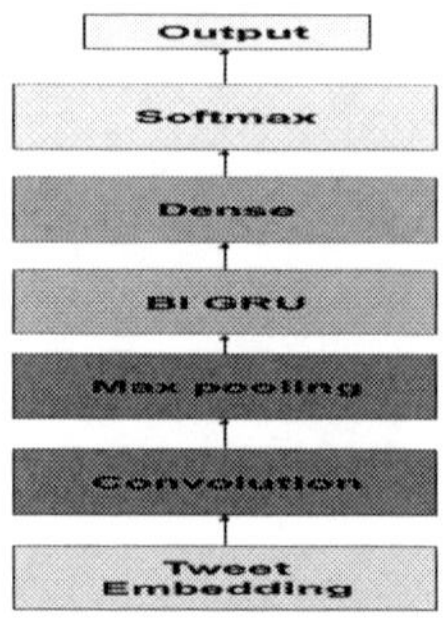

Figure 1: Single View

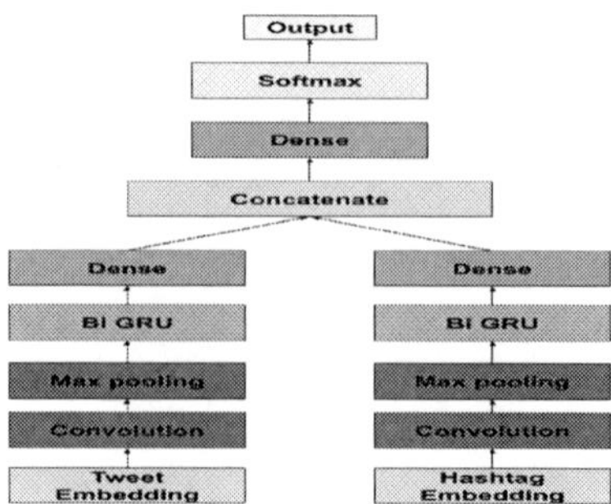

Figure 2: Two View

3 Experimental Evaluation

3.1 Experimental Set-up

We used Glove 300 dimension embeddings (Pennington et al., 2014) for both tweet and hashtags, maximum length of clean text and hashtags are set to 75 and 3 respectively. For convolution layer we used 256 filters with kernel size of 3 and pool size of 2. GRU's use less training parameters and therefore use less memory, execute faster and train faster than LSTM's, due to this we used GRU's for extracting semantic representation, we used 256 units and recurrent dropout of 0.1. We used regex to remove all unwanted symbols and to identify hashtags. Some of the lines of test data are corrupted due to which length is more than 280 for those we assigned length as 0.

	TRAIN	DEV	TEST
max len of tweet	62	61	77
avg no. of hashtags in a tweet	2.10	2.18	2.20
no. of tweets with atleast one hashtag	4277	1086	1303
total number of tweets	14717	3680	4372

Table 1: Dataset Statistics

3.2 Results

Two show the efficacy of our proposed approach, results on validation data are shown in the Table 2. From the Table 2 it is evident that our Two View system performed better than Single view system. Performance of our Two view system on the test set achieved a micro-averaged F-score of **0.58**, a micro-averaged precision of **0.54**, and a micro-averaged recall of **0.64**.

	P	R	F1
Single View	**0.52**	0.63	0.57
Two View	0.50	**0.78**	**0.61**

Table 2: Results on Validation set

4 Conclusion

Our main goal is to show the potentiality of our Two View framework which employs semantic views of both tweet and hashtag. Our approach work better when the percentage of tweets having atleast one hashtag are high. We also expect to improve the performance of our system by employing intra-attention between two views and through parameter tuning.

References

Timothy Baldwin, Marie-Catherine de Marneffe, Bo Han, Young-Bum Kim, Alan Ritter, and Wei Xu. 2015. Shared tasks of the 2015 workshop on noisy user-generated text: Twitter lexical normalization and named entity recognition. In *Proceedings of the Workshop on Noisy User-generated Text*, pages 126–135.

Mark Dredze, Nicholas Andrews, and Jay DeYoung. 2016. Twitter at the grammys: A social media corpus for entity linking and disambiguation. In *Proceedings of The Fourth International Workshop on Natural Language Processing for Social Media*, pages 20–25.

Ari Z Klein, Abeed Sarker, Davy Weissenbacher, and Graciela Gonzalez-Hernandez. 2019. Towards scaling twitter for digital epidemiology of birth defects. *NPJ digital medicine*, 2(1):1–9.

Efthymios Kouloumpis, Theresa Wilson, and Johanna Moore. 2011. Twitter sentiment analysis: The good the bad and the omg! In *Fifth International AAAI conference on weblogs and social media*.

Jeffrey Pennington, Richard Socher, and Christopher D Manning. 2014. Glove: Global vectors for word representation. In *Proceedings of the 2014 conference on empirical methods in natural language processing (EMNLP)*, pages 1532–1543.

Chanchal Suman, Saichethan Miriyala Reddy, Sriparna Saha, and Pushpak Bhattacharyya. 2020. Why pay more? a simple and efficient named entity recognition system for tweets. *Expert Systems with Applications*, page 114101.

Hao Wang, Dogan Can, Abe Kazemzadeh, François Bar, and Shrikanth Narayanan. 2012. A system for real-time twitter sentiment analysis of 2012 us presidential election cycle. In *Proceedings of the ACL 2012 system demonstrations*, pages 115–120. Association for Computational Linguistics.

Davy Weissenbacher, Abeed Sarker, Arjun Magge, Ashlynn Daughton, Karen O'Connor, Michael J. Paul, and Graciela Gonzalez-Hernandez. 2019. Overview of the fourth social media mining for health (SMM4H) shared tasks at ACL 2019. In *Proceedings of the Fourth Social Media Mining for Health Applications (#SMM4H) Workshop & Shared Task*, pages 21–30, Florence, Italy, August. Association for Computational Linguistics.

Identification of Medication Tweets Using Domain-specific Pre-trained Language Models

Yandrapati Prakash Babu
Department of Computer Applications
NIT Trichy, India.
prakash.babu23@gmail.com

Rajagopal Eswari
Department of Computer Applications
NIT Trichy, India.
eswari@nitt.edu

Abstract

In this paper, we present our approach for task1 of SMM4H 2020. This task involves automatic classification of tweets mentioning medication or dietary supplements. For this task, we experiment with pre-trained models like Biomedical RoBERTa, Clinical BERT and Biomedical BERT. Our approach achieves F1-score of 73.56%.

1 Introduction

In recent times, social media platforms like twitter, facebook, reddit attracted large number of internet users. The valuable information shared by internet users which also includes health related experiences is useful in many tasks including pharmacovigilance (Kalyan and Sangeetha, 2020b). User generated texts in social media are noisy with lots of slang words and misspelled words. We participate in task1 of SMM4H2020 which aims to develop a system that can identify tweets with medication or dietary supplement mentions. Example of tweet with medication or dietary supplement mention is 'It is good to take Vitamin C every day after lunch'. An example of a tweet without medication or dietary supplement mention is 'Vitamin C is good for health' (Wu et al., 2018). The main challenge in this task is that the system should be able to identify from the context of the tweet that the mention having drug or dietary supplement name is actually referring to the drug or dietary supplementary. This task is treated as binary classification and aims at training a model which can label the given tweet with 1 if it contains medication mention and 0 if there is no medication mention. The performance of models in this task is evaluated using F1-score of class 1.

Many research works (Kalyan and Sangeetha, 2020a; Subramanyam and S, 2020) show that models trained on medical text can better understand medical terms. So, the task is experimented with pre-trained models like Biomedical RoBERTa (Gururangan et al., 2020), Clinical BERT (Alsentzer et al., 2019) and Biomedical BERT (Lee et al., 2020). Biomedical BERT and Biomedical RoBERTa are obtained by further pre-training BERT (Devlin et al., 2019) and RoBERTa (Liu et al., 2019) models on biomedical corpus while Clinical BERT is obtained by further pre-training BERT model on MIMIC-III corpus (Johnson et al., 2016). Among these, the model based on Biomedical RoBERTa achieves the highest F1-score of 73.56%.

2 Dataset and preprocessing

The organizers of this task released train, validation and test sets. The train set includes 55419 tweets (146 positive tweets and 55273 negative tweets), validation set includes 13853 tweets (35 positive tweets and 13818 negative tweets) and test set consists of 29687. As tweets are noisy in nature, the following basic steps are used to clean the tweets:

- User mentions and urls are replaced with $< user >$ and $< url >$ respectively.

- HTML characters and unnecessary punctuation symbols are removed.

Proceedings of the 5th Social Media Mining for Health Applications (#SMM4H) Workshop & Shared Task, pages 128–130
Barcelona, Spain (Online), December 12, 2020.

Model	Precision	Recall	F1-score
Biomedical RoBERTa	65.98	83.12	**73.56**
Biomedical BERT	-	-	71.00
Clinical BERT	-	-	67.00
Average score of all Task-1 teams	70.32	69.48	66.28

Table 1: Precision, Recall and F1-score of our models on test data.

- Emojis are replaced with their corresponding descriptions.

- Twitter slang words are replaced with corresponding standard words. For example, *'lol'* is replaced with *'laugh out loud'*.

3   Model Description

In recent times, researchers have focused on exploiting deep pre-trained language models like BERT, RoBERTa in most of the natural language processing tasks. These models are adapted to medical domain by means of additional training on large medical text. This paper investigates how well domain specific models like Biomedical RoBERTa, Biomedical BERT and Clinical BERT identify medication tweets. First, tweet representation $e_t \in \mathbb{R}^n$ is generated using domain specific models, n represents hidden state vector in pre-trained language model which is equal to 768. Then, sigmoid layer with parameters $W \in \mathbb{R}^{n \times 1}$ and $b \in \mathbb{R}$ is applied on e_t to transform it into single value which represents the predicted label q.

$$e_t = PTLM(tweet) \tag{1}$$

$$q = Sigmoid(W^T e_t + b) \tag{2}$$

Here PTLM refers to pretrained language model and it can be Biomedical RoBERTa, Biomedical BERT or Clinical BERT.

4   Experiments and Results

To handle imbalance in the dataset, the dataset is augmented with 9622 tweets (4975 positive tweets and 4647 negative tweets) from SMM4H 2018 task1 dataset (Weissenbacher et al., 2018). Further, positive tweets are up sampled 15 times and 90% of the negative tweets are randomly chosen. We use validation set to find optimal values for various hyperparameters. We use batch size of 16, learning rate of 3e-5 and train the model for 3 epochs. All our models are implemented using transformers library in PyTorch (Wolf et al., 2019). The task organizers released precision and recall scores only for which model got the highest F1-score, the performance of our models and average score is listed in Table 1. Among the three models, the model based on Biomedical RoBERTa outperforms other models and achieves the highest F1-score of 73.56%. As a whole, our approach achieves good results which is much higher than the average scores.

5   Conclusion

The medication mentions in tweets are identified using domain specific deep pre-trained models. Experimental results show that the model based on Biomedical RoBERTa achieves the best F1-score of 73.56% which is significantly higher than the average F1-score of 66.28%.

References

Emily Alsentzer, John Murphy, William Boag, Wei-Hung Weng, Di Jindi, Tristan Naumann, and Matthew McDermott. 2019. Publicly available clinical bert embeddings. In *Proceedings of the 2nd Clinical Natural Language Processing Workshop*, pages 72–78.

Jacob Devlin, Ming-Wei Chang, Kenton Lee, and Kristina Toutanova. 2019. Bert: Pre-training of deep bidirectional transformers for language understanding. In *Proceedings of the 2019 Conference of the North American Chapter of the Association for Computational Linguistics: Human Language Technologies, Volume 1 (Long and Short Papers)*, pages 4171–4186.

Suchin Gururangan, Ana Marasović, Swabha Swayamdipta, Kyle Lo, Iz Beltagy, Doug Downey, and Noah A Smith. 2020. Don't stop pretraining: Adapt language models to domains and tasks. *arXiv preprint arXiv:2004.10964*.

Alistair EW Johnson, Tom J Pollard, Lu Shen, H Lehman Li-Wei, Mengling Feng, Mohammad Ghassemi, Benjamin Moody, Peter Szolovits, Leo Anthony Celi, and Roger G Mark. 2016. Mimic-iii, a freely accessible critical care database. *Scientific data*, 3(1):1–9.

Katikapalli Subramanyam Kalyan and S Sangeetha. 2020a. Bertmcn: Mapping colloquial phrases to standard medical concepts using bert and highway network. Technical report, EasyChair.

Katikapalli Subramanyam Kalyan and S Sangeetha. 2020b. Secnlp: A survey of embeddings in clinical natural language processing. *Journal of biomedical informatics*, 101:103323.

Jinhyuk Lee, Wonjin Yoon, Sungdong Kim, Donghyeon Kim, Sunkyu Kim, Chan Ho So, and Jaewoo Kang. 2020. Biobert: a pre-trained biomedical language representation model for biomedical text mining. *Bioinformatics*, 36(4):1234–1240.

Yinhan Liu, Myle Ott, Naman Goyal, Jingfei Du, Mandar Joshi, Danqi Chen, Omer Levy, Mike Lewis, Luke Zettlemoyer, and Veselin Stoyanov. 2019. Roberta: A robustly optimized bert pretraining approach. *arXiv preprint arXiv:1907.11692*.

Kalyan Katikapalli Subramanyam and Sangeetha S. 2020. Deep contextualized medical concept normalization in social media text. *Procedia Computer Science*, 171:1353 – 1362. Third International Conference on Computing and Network Communications (CoCoNet'19).

Davy Weissenbacher, Abeed Sarker, Michael Paul, and Graciela Gonzalez. 2018. Overview of the third social media mining for health (smm4h) shared tasks at emnlp 2018. In *Proceedings of the 2018 EMNLP Workshop SMM4H: The 3rd Social Media Mining for Health Applications Workshop & Shared Task*, pages 13–16.

Thomas Wolf, Lysandre Debut, Victor Sanh, Julien Chaumond, Clement Delangue, Anthony Moi, Pierric Cistac, Tim Rault, Rémi Louf, Morgan Funtowicz, et al. 2019. Huggingface's transformers: State-of-the-art natural language processing. *ArXiv*, pages arXiv–1910.

Chuhan Wu, Fangzhao Wu, Junxin Liu, Sixing Wu, Yongfeng Huang, and Xing Xie. 2018. Detecting tweets mentioning drug name and adverse drug reaction with hierarchical tweet representation and multi-head self-attention. In *Proceedings of the 2018 EMNLP Workshop SMM4H: The 3rd Social Media Mining for Health Applications Workshop & Shared Task*, pages 34–37.

Medication Mention Detection in Tweets
Using ELECTRA Transformers and Decision Trees

Lung-Hao Lee*, Po-Han Chen, Hao-Chuan Kao
Ting-Chun Hung, Po-Lei Lee and Kuo-Kai Shyu
Department of Electrical Engineering, National Central University, Taiwan
Pervasive Artificial Intelligence Research (PAIR) Labs, Taiwan
`*lhee@ee.ncu.edu.tw`

Abstract

This study describes our proposed model design for the SMM4H 2020 Task 1. We fine-tune ELECTRA transformers using our trained SVM filter for data augmentation, along with decision trees to detect medication mentions in tweets. Our best F1-score of 0.7578 exceeded the mean score 0.6646 of all 15 submitting teams.

1 Introduction

The Social Media Mining for Health Applications (SMM4H) shared task involves natural language processing challenges for using social media data for health research. We participated in the SMM4H 2020 Task 1, focusing on automatic classification of tweets that mention medications (Klein et al., 2020). This binary classification task involves distinguishing tweets that mention a medication or dietary supplement (annotated as '1') from those that do not (annotated as '0'). This task was first organized in 2018 using a data set containing an artificially balanced distribution of the two classes (Weissenbacher et al., 2018). Several approaches have been presented to address this binary classification task (Coltekin and Rama, 2018; Xherija, 2018; Wu et al., 2018). However, this year's task is more challenging. The data set represents a natural, highly imbalanced distribution of the two classes from tweets posted by 112 women during pregnancy, with only approximately 0.2% of the tweets mentioning a medication (Sarker et al., 2017; Weissenbacher et al., 2019).

This paper describes the NCUEE (National Central University, Dept. of Electrical Engineering) system for the SMM4H 2020 Task 1. To deal with highly imbalanced distribution, the support vector machine trained using the training data is used as a filter to crawl and select more tweets for data augmentation. We then fine-tune the pre-trained ELECTRA transformers (Clark et al., 2020), using our augmented data for medication mention detection. In addition, we train the decision tree as a supplementary classifier. Finally, the integrated set of testing instances are detected as a positive class from ELECTRA and decision trees are regraded as label '1' and the remaining cases as label '0' to form our submissions.

2 The NCUEE System

In addition to the training data provided by task organizers, we crawl and select highly related tweets for data augmentation. We manually check small positive tweets to pick up textual terms that may refer to medications, and then use these terms as seeds for query expansions. The pre-trained Word2Vec embedding from Twitter data is used to look up word vectors and compare their cosine similarities. The top 10 similar terms of seeds are collected, where expanded terms are kept if the document frequency (DF) of an expanded term in the positive class exceeds that in the negative class. Each expanded term, along with the query term 'pregnant' is regarded as an individual query to search for possibly related tweets from Twitter. To automatically label highly positive cases, we train the support vector machine (SVM)

Proceedings of the 5th Social Media Mining for Health Applications (#SMM4H) Workshop & Shared Task, pages 131–133
Barcelona, Spain (Online), December 12, 2020.

using the provided training data and select crawled tweets predicted to be positive cases by the SVM. Finally, we construct an augmented data set including the original training sets for neural computing.

ELECTRA (Efficiently Learning as Encoder that Classifiers Token Replacements Accurately) is a new pre-training approach that aims to match or exceed the downstream performance of an MLM (Masked Language Modeling) pre-trained model while using less computational loading (Clark et al., 2020). ELECTRA trains two transformer models: the generator, which replaces the tokens in a sequence for training a masked language model; and the discriminator, which tries to identify which tokens in the sequence were replaced by the generator. We use pre-trained ELECTRA transformers and fine-tune them using our augmented data to detect medication mentions in tweets.

According to our empirical results from the validation set, the decision tree (DT) classifiers usually achieved a high degree of precision if the discriminated features had been extracted and learned, but very low recall if the testing cases were significantly different from the trained ones. Hence, we use the same trained SVM as a filter to select the positive cases (predicted as '1') of from the 2018 task training data that may be closely similar to the positive tweets in this task and include these in an augmented set. We then adopt the TF-IDF (Term Frequency-Inverse Document Frequency) weighting method to extract discriminated features of positive tweets from this augmented set and use them with the original training data to train the decision trees.

Finally, based on our error analysis, we found the decision tree classifier was partially complementary to ELECTRA. So, the integrated set of testing instances predicted as the positive class from the ELEC-TRA transformers and decision trees are labeled '1', otherwise '0' in our submissions.

3   Evaluation

We picked up 112 seeds from 181 positive tweets to further expand the dataset by 72 unique terms via cosine similarity through the pre-trained Word2Vec embeddings of the tweets. Without using the SVM as a filter, we have 57,678 positive tweets and 105,273 negative tweets. With SVM, we have 32,619 positive tweets and 65,238 negative tweets. The distribution of tweets after data augmentation (DA) was still remained imbalanced, with a positive to negative ratio close to 1:2. The pre-trained ELECTRA-Large was downloaded from HuggingFace (Wolf et al., 2019). The hyper-parameters used for fine-tuning ELECTRA are as follows: batch size 16; gradient accumulation steps 16; learning rate 1e-5; and number of training epochs 6.

Table 1 shows the results on the validation and test sets. The evaluation metric is the F1-score for the positive class (i.e. tweets that mention medications). For the test set, compared with submission 1 that does not use SVM as a filter for data augmentation and submission 3 that adds the prediction result of the RoBERTa transformer (Liu et al., 2019), our submission 2 achieved the best F1-score of 0.7578. Their relative ranks were identical to those of the validation set.

#	Methods	Validation Set			Test Set		
		P	R	F1	P	R	F1
1	without SVM, ELECTRA + DT	0.9412	0.9143	0.9275	0.7910	0.6883	0.7361
2	with SVM, ELECTRA + DT	0.9444	0.9714	**0.9577**	0.7262	0.7922	**0.7578**
3	with SVM, ELECTRA + DT+ RoBERTa	0.8750	1.0000	0.9333	0.6702	0.8182	0.7368

Table 1. Submission evaluation results.

4   Conclusions

This study describes the NCUEE system participating in the SMM4H 2020 Task 1 for medication mention detection, including system design, implementation and evaluation. Our best F1-score of 0.7578 exceeded the mean score 0.6646 for all 15 teams with at least one submissions.

Acknowledgements

This study is partially supported by the Ministry of Science and Technology, Taiwan under the grant MOST 108-2218-E-008-017-MY3 and MOST 108-2634-F-008-003- through Pervasive Artificial Intelligence Research (PAIR) Labs, Taiwan.

Reference

Abeed Sarker, Pramod Chandrashekar, Arjun Magge, Haitao Cai, Ari Klein and Graciela Gonzalez. 2017. Discovering cohorts of pregnant women from social media for safety surveillance analysis. *Journal of Medical Internet Research*, 19(10):e361.

Ari Z. Klein, Ilseyar Alimova, Ivan Flores, Arjun Magge, Zulfat Miftahutdinov, Anne-Lyse Minard, Karen O'Connor, Abeed Sarker, Elena Tutubalina, Davy Weissenbacher, and Graciela Gonzalez-Hernandez. 2020. Overview of the fifth Social Media Mining for Health Applications (#SMM4H) Shared Tasks at COLING 2020. In *Proceedings of the Fifth Social Media Mining for Health Applications (#SMM4H) Workshop & Shared Task*.

Cagri Coltekin and Taraka Rama. 2018. Drug-use identification from tweets with word and character n-grams. In *Proceedings of the 3rd Social Media Mining for Health Applications (SMM4H) Workshop and Shared Task*, pages 52-53, Brussels, Belgium.

Chuhan Wu, Fangzhao Wu, Junxin Liu, Sixing Wu, Yongfeng Huang and Xing Xie. 2018. Detecting tweets mentioning drug name and adverse drug reaction with hierarchical tween representation and multi-head self-attention. In *Proceedings of the 3rd Social Media Mining for Health Applications (SMM4H) Workshop and Shared Task*, pages 34-37, Brussels, Belgium.

Davy Weissenbacher, Abeed Sarker, Ari Klein, Karen O'Connorm Arjun Magge and Graciela Gonzalez-Hernandez. 2019. Deep neural networks ensemble for detecting medication mentions in tweets. *Journal of the American Medical Informatics Association*, 26(12):1618-1626.

Davy Weissenbacher, Abeed Sarker, Michael Paul and Graciela Gonzalez-Hernandez. 2018. Overview of the third social media mining for health (SMM4H) shared tasks at EMNLP 2018. In *Proceedings of the 3rd Social Media Mining for Health Applications (SMM4H) Workshop and Shared Task*, pages 13-16, Brussels, Belgium.

Kevin Clark, Minh-Thang Luong, Quoc V. Le and Christopher D. Manning. 2020. ELECTRA: pre-training text encoders as discriminators rather than generators. In *Proceedings of the 8th International Conference on Learning Representations (ICLR)* , pages 1-18.

Orest Xherija. 2018. Classification of medication-related tweets using stacked bidirectional LSTMs with context-aware attention. In *Proceedings of the 3rd Social Media Mining for Health Applications (SMM4H) Workshop and Shared Task*, pages 38-42, Brussels, Belgium.

Thomas Wolf, Lysandre Debut, Victor Sanh, Julien Chaumond, Clement Delangue, Anthony Moi, Pierric Cistac, Tim Rault, Remi Louf, Morgan Funtowicz, Joe Davison, Sam Shleifer, Patrick von Platen, Clara Ma, Yacine Jernite, Julien Plu, Canwen Xu, Teven Le Scao, Sylvain Gugger, Mariama Drame, Quentin Lhoest, and Alexander M. Rush. 2019. HuggingFace's transformers: state-of-the-art natural language processing, *arXiv:1910.03771*.

Yinhan Liu, Myle Ott, Naman Goyal, Jingfei Du, Mandar Joshi, Danqi Chen, Omer Levy, Mike Lewis, Luke Zettlemoyer and Veselin Stoyanov. 2019. RoBERTa: a robustly optimized BERT pretraining approach. *Computing Research Repository*, *arXiv:1907.11692*. version 2.

LITL at SMM4H: an old-school feature-based classifier for identifying adverse effects in Tweets

Ludovic Tanguy, Lydia-Mai Ho-Dac, Cécile Fabre
and the **Master LITL students**: Roxane Bois, Touati Mohamed Yacine Haddad,
Claire Ibarboure, Marie Joyau, François Le moal, Jade Moillic,
Laura Roudaut, Mathilde Simounet, Irena Stankovic, Mickaela Vandewaetere

CLLE: CNRS & University of Toulouse, France
{ludovic.tanguy, lydia-mai.ho-dac, cecile.fabre}@univ-tlse2.fr

Abstract

This paper describes our participation to the SMM4H shared task 2. We designed a linear classifier that estimates whether a tweet mentions an adverse effect associated to a medication. Our system addresses English and French, and is based on a number of ad-hoc word lists and features. These cues were mostly obtained through an extensive corpus analysis of the provided training data. Different weighting schemes were tested (manually tuned or based on a logistic regression), the best one achieving a F1 score of 0.31 for English and 0.15 for French.

1   Overview

This article describes the participation of the students of the *LITL* master and their teachers to the Social Media Mining for Health (SMM4H) shared task 2 (Klein et al., 2020). LITL (stands for *Linguistique, Informatique, Technologies du Langage*, i.e. Linguistics, IT, Language technologies) is a master's program at the University of Toulouse, France that is mainly aimed at linguistics and humanities students.

The shared task is a binary classification of Twitter messages in different languages, indicating whether the message contains a mention of medication adverse effects. Participation to this task was part of the first year students' curriculum. At this stage, their computer skills were still limited to corpus processing and simple programs, so it was decided that the system's architecture would be a traditional linear classifier based on ad-hoc features. This approach was also deemed justified given the heavily biased distribution of data (known to be an issue for most machine learning techniques). However, the students were encouraged to apply and hone their corpus linguistics skills, and to perform some feature engineering. The approach was the following:

1. Observe the training data with corpus analysis tools, in order to identify the main characteristics of the target (i.e. tweets evoking an adverse effect);
2. Build word lists and design simple features for each of these characteristics;
3. Design a program that computes the features' values on the target data and implements a simple weight-based linear classifier;
4. Tune the weights in order to maximize the classifier's performance on the validation data.

Due to the necessity to actually observe and understand the training data only the French and English sets were considered, as none of the students was proficient in Russian.

2   Technical details

For observation and actual processing in both languages the tweets were preprocessed as follows:

- retweet marks (`rt @X`) were removed;
- user names (`@XXX`) were replaced with a generic and POS-wide unambiguous proper name (*Sacha*);
- URLs and email addresses were replaced with generic placeholders (`<URL/>` and `<email/>`);
- non-standard spelling was normalised (e.g. removal of exceeding repeated letters *baaad* → *bad*);
- POS tagging and lemmatizing were performed, using the Talismane toolkit for both target languages (Urieli, 2013).

Proceedings of the 5th Social Media Mining for Health Applications (#SMM4H) Workshop & Shared Task, pages 134–137
Barcelona, Spain (Online), December 12, 2020.

Fifteen word lists were compiled for each language, each one targeting a specific aspect of the tweets content. Those word lists contain keywords extracted from the target tweets and non target tweets using the TXM corpus analysis tool (Heiden et al., 2010). The lists were extended with existing lexical resources such as sentiment lexicons (e.g. the SocialSent lexicon (Abdaoui et al., 2017) and the FEEL – French Expanded Emotion Lexicon (Hamilton et al., 2016)) and biomedical domain language resources (Névéol et al., 2014). Table 1 gives an overview of the word lists designed and used as features for English and French Tweet classification.

Target tweet keywords (i.e. positively correlated with adverse effect)			
Word list	**Example (English)**	**# items (English)**	**# items (French)**
Symptoms	*headache, cough, addict...*	378	376
Causal verbs	*impact, stop...*	41	58
Sentiment (negative)	*dirty, resent...*	3647	894
Body parts	*chest, joint...*	86	84
Medication (first set)	*Effexor, Paxil...*	21	22
Increase verbs	*gain, raise...*	52	112
Decrease verbs	*decline, reduce...*	38	107
First person pronouns and determiners	*I, our*	7	12
Negation	*not, cannot...*	9	15
Emojis (negative)	☹, 😣	23	23
Non target tweet keywords (i.e. negatively correlated with adverse effect)			
Word list	**Example (English)**	**# items (English)**	**# items (French)**
Misc. verbs	*approve, study...*	18	8
Sentiments (positive)	*great, secure...*	2080	848
Medication (second set)	*Floxin, Prozac...*	40	9
2nd and 3rd person pronouns and determiners	*you, himself*	17	22
Emojis (positive)	☺	57	57

Table 1: Word lists designed and used as features

Each word list led to a numeric feature corresponding to the raw frequency of matching lemmas in the tweet. Two different strategies were considered for dealing with multi-word expressions. Runs 1 and 3 count all items as matches even in case they are also part of an item in another list, e.g. *skin* (body part) and *skin rash* (symptom). In contrast, run 2 only counts the longer item (e.g. *skin rash* as a symptom feature).

Three additional non-lexical features were also used: number of hash tags, number of URLs and number of Twitter user names (i.e. *Sacha*, cf. *supra*).

Each feature was assigned a weight proportional to its relative importance in the decision process. For runs 2 and 3 the weights were individually fixed based on the frequency ratios in target (*vs* non target) tweets in the training data, and then manually adjusted based on the scores obtained on the validation sets. The best weights were found by progressively increasing the weight of each feature independently of each other until the best F1 score is reached. For run 2, we adopted a principle of equality between features. For run 3, some features were considered as more important than others on the basis of manual observations. Run 1 used a standard logistic regression classifier trained on training data.

3   Results and discussion

Table 2 shows the results for each run and for each language on the validation and test sets. The first strategy for dealing with multi-word expressions was clearly better. Manual tuning of the feature weights (which was performed before the students were introduced to machine learning techniques) was promising on the validation set (especially regarding precision) but proved to be much less robust in the test set. Further experiments will be performed in order to assess the added value of selected word lists, compared to more straightforward and non-selective bag-of-words methods, and of course more recent NLP techniques based on word embeddings and neural classifiers.

Language	Run	Validation			Test
		R	P	F1	F1
English	1 (logistic regression, all items)	0.40	0.16	0.23	**0.31**
	2 (adjusted, longer items only)	**0.55**	0.23	0.32	0.25
	3 (adjusted, all items)	0.24	**0.56**	**0.33**	0.27
French	1 (logistic regression, all items)	**1.00**	0.13	0.22	**0.15**
	2 (adjusted, longer items only)	0.50	0.13	0.20	0.00
	3 (adjusted, all items)	0.33	**0.18**	**0.23**	0.12

Table 2: Results

References

Amine Abdaoui, Jérôme Azé, Sandra Bringay, and Pascal Poncelet. 2017. Feel: a French expanded emotion lexicon. *Language Resources and Evaluation*, 51(3):833–855.

Elise Bigeard, Natalia Grabar, and Frantz Thiessard. 2018. Detection and analysis of drug misuses. a study based on social media messages. *Frontiers in pharmacology*, 9:791.

Tilia Ellendorff, Joseph Cornelius, Heath Gordon, Nicola Colic, and Fabio Rinaldi. 2018. UZH@SMM4H: System descriptions. In *Proceedings of the 2018 EMNLP Workshop SMM4H: The 3rd Social Media Mining for Health Applications Workshop & Shared Task*, pages 56–60.

Rachel Ginn, Pranoti Pimpalkhute, Azadeh Nikfarjam, Apurv Patki, Karen O'Connor, Abeed Sarker, Karen Smith, and Graciela Gonzalez. 2014. Mining Twitter for adverse drug reaction mentions: a corpus and classification benchmark. In *Proceedings of the fourth workshop on building and evaluating resources for health and biomedical text processing*, pages 1–8.

William L Hamilton, Kevin Clark, Jure Leskovec, and Dan Jurafsky. 2016. Inducing domain-specific sentiment lexicons from unlabeled corpora. In *Proceedings of the Conference on Empirical Methods in Natural Language Processing. Conference on Empirical Methods in Natural Language Processing*, volume 2016, page 595.

Serge Heiden, Jean-Philippe Magué, and Bénédicte Pincemin. 2010. TXM : Une plateforme logicielle open-source pour la textométrie – conception et développement. In Sergio Bolasco, Isabella Chiari, and Luca Giuliano, editors, *10th International Conference on the Statistical Analysis of Textual Data - JADT 2010*, volume 2, pages 1021–1032, Rome, Italy. Edizioni Universitarie di Lettere Economia Diritto.

Ferdaous Jenhani, Mohamed Salah Gouider, and Lamjed Ben Said. 2016. A hybrid approach for drug abuse events extraction from Twitter. *Procedia computer science*, 96:1032–1040.

Ferdaous Jenhani, Mohamed Salah Gouider, and Lamjed Ben Said. 2019. Hybrid system for information extraction from social media text: Drug abuse case study. *Procedia Computer Science*, 159:688–697.

Svetlana Kiritchenko, Saif M Mohammad, Jason Morin, and Berry de Bruijn. 2017. NRC-Canada at SMM4H shared task: Classifying tweets mentioning adverse drug reactions and medication intake. In *SMM4H@ AMIA*.

Ari Z. Klein, Ilseyar Alimova, Ivan Flores, Arjun Magge, Zulfat Miftahutdinov, Anne-Lyse Minard, Karen O'Connor, Abeed Sarker, Elena Tutubalina, Davy Weissenbacher, and Graciela Gonzalez-Hernandez. 2020. Overview of the fifth social media mining for health applications (#smm4h) shared tasks at coling 2020. In *Proceedings of the Fifth Social Media Mining for Health Applications (#SMM4H) Workshop & Shared Task*.

Alex Lamb, Michael Paul, and Mark Dredze. 2013. Separating fact from fear: Tracking flu infections on Twitter. In *Proceedings of the 2013 Conference of the North American Chapter of the Association for Computational Linguistics: Human Language Technologies*, pages 789–795.

Anne-Lyse Minard, Christian Raymond, and Vincent Claveau. 2018. Participation de l'IRISA à DeFT 2018 : classification et annotation d'opinion dans des tweets. In *Actes de la conférence Traitement Automatique de la Langue Naturelle, TALN 2018*, page 265.

Aurélie Névéol, Julien Grosjean, Stéfan Jacques Darmoni, Pierre Zweigenbaum, et al. 2014. Language resources for French in the biomedical domain. In *Proceedings of LREC*, pages 2146–2151.

Karen O'Connor, Pranoti Pimpalkhute, Azadeh Nikfarjam, Rachel Ginn, Karen L Smith, and Graciela Gonzalez. 2014. Pharmacovigilance on Twitter? Mining tweets for adverse drug reactions. In *AMIA annual symposium proceedings*, volume 2014, page 924. American Medical Informatics Association.

Abeed Sarker and Graciela Gonzalez-Hernandez. 2017. Overview of the second social media mining for health (SMM4) shared tasks at amia 2017. *Training*, 1(10,822):1239.

Abeed Sarker and Graciela Gonzalez. 2015. Portable automatic text classification for adverse drug reaction detection via multi-corpus training. *Journal of biomedical informatics*, 53:196–207.

Abeed Sarker, Azadeh Nikfarjam, and Graciela Gonzalez. 2016. Social media mining shared task workshop. In *Biocomputing 2016: Proceedings of the Pacific Symposium*, pages 581–592.

Assaf Urieli. 2013. *Robust French syntax analysis: reconciling statistical methods and linguistic knowledge in the Talismane toolkit*. Phd thesis, University of Toulouse, France.

Davy Weissenbacher, Abeed Sarker, Michael Paul, and Graciela Gonzalez. 2018. Overview of the third social media mining for health (SMM4H) shared tasks at EMNLP 2018. In *Proceedings of the 2018 EMNLP Workshop SMM4H: The 3rd Social Media Mining for Health Applications Workshop & Shared Task*, pages 13–16.

Davy Weissenbacher, Abeed Sarker, Arjun Magge, Ashlynn Daughton, Karen O'Connor, Michael Paul, and Graciela Gonzalez. 2019. Overview of the fourth social media mining for health (SMM4H) shared tasks at ACL 2019. In *Proceedings of the Fourth Social Media Mining for Health Applications (# SMM4H) Workshop & Shared Task*, pages 21–30.

Theresa Wilson, Janyce Wiebe, and Paul Hoffmann. 2005. Recognizing contextual polarity in phrase-level sentiment analysis. In *Proceedings of human language technology conference and conference on empirical methods in natural language processing*, pages 347–354.

Sentence Classification with Imbalanced Data for Health Applications

Farhana Ferdousi Liza
School of Computer Science and Electronic Engineering
University of Essex, Colchester, UK
farhana.ferdousi.liza@essex.ac.uk

Abstract

Identifying and extracting reports of medications, their abuse or adverse effects from social media is a challenging task. In social media, relevant reports are very infrequent, causes imbalanced class distribution for machine learning algorithms. Learning algorithms typically designed to optimize the overall accuracy without considering the relative distribution of each class. Thus, imbalanced class distribution is problematic as learning algorithms have low predictive accuracy for the infrequent class. Moreover, social media represents natural linguistic variation in creative language expressions. In this paper, we have used a combination of data balancing and neural language representation techniques to address the challenges. Specifically, we participated the shared tasks 1, 2 (all languages), 4, and 3 (only the span detection, no normalization was attempted) in Social Media Mining for Health applications (SMM4H) 2020 (Klein et al., 2020). The results show that with the proposed methodology recall scores are better than the precision scores for the shared tasks. The recall score is also better compared to the mean score of the total submissions. However, the F1-score is worse than the mean score except for task 2 (French).

1 Introduction

Advances in representation learning that attempts to automatically learn features for natural language processing (Young et al., 2018) present the possibility of utilizing social media (i.e. Twitter) data source for public health applications such as health monitoring and surveillance. Several ethical, legal and methodological challenges need to be addressed that are unique to Twitter data source (Ahmed et al., 2017). The ongoing shared tasks in Social Media Mining for Health applications (SMM4H) define evolving challenges specific to the Twitter data source for health domain (Weissenbacher et al., 2019). To address the methodological challenges, in recent years, several techniques have been proposed based on the SMM4H shared tasks (Sarker et al., 2018; Weissenbacher et al., 2019).

Epidemiologists intend to detect mentions of health issues related to medications early on Twitter. The adverse effect of medications is one of the leading causes of post-therapeutic deaths (Saha et al., 2018). One of the challenges of detecting real reports on medications, their abuse or adverse effects is to distinguish the relevant true reports from other general statements, news, and institutional advice. Reports on health issues (i.e. abuse or adverse effects of a medication) in social media are rare instances (i.e. very small percentage contain relevant information) (Weiss, 2004; Batista et al., 2004). This causes a major machine learning methodological challenges called imbalanced learning problem (He and Garcia, 2009; Ling and Sheng, 2010). The learning problem happens in the presence of underrepresented data and severely skewed class distribution, the situation where one of the class categories comprises a significantly larger proportion of the dataset than the other classes. Imbalanced class distribution is a practical issue in most real-world datasets (e.g. fraud detection, disease detection) and complicates learning when the identification of the minority class is of specific importance. This is a general problem for health domain with medical data utilizing machine learning methods (Rahman and Davis, 2013).

Proceedings of the 5th Social Media Mining for Health Applications (#SMM4H) Workshop & Shared Task, pages 138–145
Barcelona, Spain (Online), December 12, 2020.

When dealing with imbalanced data, model evaluation is different than typical class-balanced loss based machine learning framework (Cui et al., 2019; Marchand and Strawderman, 2020). Models that trained by minimizing errors on imbalanced datasets, tend to frequently predict the majority class; achieving high overall accuracy in such cases can be misleading. As class imbalance is a widespread issue, multiple techniques have been developed that help alleviate the issue (Buda et al., 2018; Haixiang et al., 2017), by either adjusting the model (e.g. changing the performance metric) or changing the data distribution (e.g. oversampling the minority class or undersampling the majority class). In this paper, we have explored novel strategies to handle extreme imbalanced class distribution.

Another challenge is representing informal Twitter text efficiently for health monitoring. Tweets contain misspelled unnormalized health concepts, expressed in a noisy, ungrammatical, multilingual, and ambiguous way. Moreover, creative and colloquial language expressions are prevalent in Twitter text. In this paper, advanced preprocessing and feature learning techniques were utilized to efficiently capture syntactic and semantic regularities of the substantially informal Twitter data. Overall, several techniques were explored to address challenges in Twitter data source for health monitoring and surveillance. The paper is organized as follows: section 2 describes data and problem statement, section 3 and 4 include the methodology used to model the datasets, and section 5 contains a discussion of the obtained results to understand the challenges specifically related to imbalanced dataset.

2   Problem Statement and Data Description

In this paper, six datasets from the Social Media Mining for Health application (SMM4H) 2020 Shared tasks were used for health monitoring and surveillance challenges. All the shared task datasets are labeled and can be modeled in one of three popular types of supervised classification problems. Binary (B) classification is the task of classifying the elements of a given set into two groups whereas multi-class (M) classification generalizes into more than two groups. Span detection (D) and normalization (N) task is defined as D+N which can be modeled as multi-class classification task followed by named-entity recognition modeling (Nadeau and Sekine, 2007; Lample et al., 2016).

Task ID: description	Type	IR (%)	# (*)	#Train	#Valid.	#Test
T1: Med. Mention	B	0.26	181 (1) 69,091 (0)	55419	13853	29687
T2 (En): AE in English	B	9.25	2,374 (1) 23,298(0)	20544	5134	4759
T2 (Fr): AE in French	B	1.61	39(1) 2,387(0)	1941	485	607
T2 (Ru): AE in Russian	B	8.75	666 (1) 6,946(0)	6090	1522	1903
T3: AE	D+N	51.20	1,212 (1) 1,155(0)	2246	560	976
T4: Med. Abuse	M	15.99	1685('a') 5488('m') 2940('c') 424('u')	10537	2635	3271

Table 1: Brief data description: rows correspond to the datasets and columns correspond to various attributes related to the datasets. The first column describes the task with an identifier. The second column denotes a classification type for the task (i.e. binary, multi-class and span-detection). The third column describes the imbalance ratio (IR), defined as the ratio of the minority class examples to the total number of examples, in percentage. The fourth column (# (*)) has the class/label distribution over the training dataset where #(*) denotes the number of Tweets labeled with *. Each dataset is comprised of three splits: training, validation and testing set. The fifth (i.e. #Train), sixth (i.e. #Valid.) and seventh (i.e. #Test) column correspond to the number of Tweets available for training, validation and testing set.

Table 1 shows a brief description of the datasets from the SMM4H 2020 shared tasks (Klein et al., 2020). Task 1 requires distinguishing between two classes of Tweets and modeled as a binary classification task. Positive Tweets that mention a medication or dietary supplement are labeled as "1" and negative Tweets that do not mention are labeled as "0". The dataset consists of Tweets posted by 112 women during pregnancy, with approximately 0.26% (see column IR%) of the training Tweets mentioning a medication or dietary supplement. The data set represents an extremely[1] imbalanced class distribution.

Task 2 is another binary classification task that involves classifying Tweets based on the mentions of adverse effect (AE) of a medication. This task includes distinct sets of Tweets posted in three languages: English (En), French (Fr) and Russian (Ru). In Tab. 1, rows corresponding to task id T2 (En), T2 (Fr) and T2 (Ru) refer to sub-tasks in three languages. Tweets that mention an adverse effect of medications are labeled as "1" and those that do not have mention are labeled as "0". The tasks require taking into account subtle linguistic and semantic variations between AEs and indications (i.e. the reason for using the medication). The T2 (Fr), T2 (En) and T2 (Ru) datasets represent moderately imbalanced class distribution with IR of 1.61%, 9.25% and 8.75% respectively.

Task 3 involves detecting the span of Tweet containing an adverse effect (AE) of a medication and then mapping the extracted AE to a standard concept identifier (ID) in the MedDRA vocabulary (preferred terms). The training data includes Tweets that report or indicate an AE that are labeled as "1" and those that do not mention are labeled as "0". The detection task thus requires a model to distinguish between AEs and indications. The class distribution of the dataset is balanced (IR=51.20%) for the detection task. The normalization task involves classification to multiple classes where each class can be defined as a standard MedDRA ID. We have not attempted the normalization part of the task and left for future work.

Task 4 requires distinguishing between more than two classes of Tweets and can be modeled as a multi-class classification problem. The task involves distinguishing among Tweets that mention at least one prescription opioid, benzodiazepine, atypical anti-psychotic, central nervous system stimulant or GABA analogue. Tweets that report potential abuse/misuse are labeled as "A" from those that report non-abuse/-misuse consumption which are labeled as "C", merely mention of the medication are labeled as "M", and unrelated are labeled as "U". The task has moderately imbalanced class distribution with IR of 15.99%.

3 Methodology for Classification Tasks

We have developed systems for shared tasks 1, 2 (all languages), 4, and 3 (only the span detection, no normalization was attempted). The main focus was towards the imbalanced classification tasks. In this section, we will describe the methods used for five classification tasks and in the next section, we will describe the method developed for the span-detection part of task 3. The models were trained on the training dataset and evaluated on validation dataset; test data was not available when the methods were being developed. Each solution of the binary and multi-class classification task can be subdivided into four common steps and these steps will be elaborated in the following subsections.

3.1 Preprocessing

Tweet language processing is challenging compared to the standard text found in the news, journals and books (Balahur, 2013). For example, Twitter users use an informal language (Tan et al., 2015) that uses special expressions, such as "lol", "omg", emoticons (Derks et al., 2008), and emphasize or exaggerate the underlying meaning of the root word by using stretched words like 'heellllp' or 'heyyyyy'. Using the stretched word is common in spoken language but Tweets can have them in written format. The traditional mainstream natural language processing tools were not designed to include the syntactic regularities of informal language (Kong et al., 2014). Additionally, Twitter language has specific metadata, such as "RT" defines Tweets that reposted by other users, the markup of topics using the hashtag ("#") sign and defining other Twitter users by "@" sign. In this paper, we have considered the Tweet specific characteristics when preprocessing the Tweets so that the language can be normalized — converting them into a more standard form of language — efficiently. Normalization of Twitter posts enable us to apply standard natural language processing (NLP) techniques more effectively.

---

[1]https://developers.google.com/machine-learning/data-prep/construct/sampling-splitting/imbalanced-data

A basic normalization (Kaufmann and Kalita, 2010; Declerck and Lendvai, 2015; Beckley, 2015) was employed as part of advanced preprocessing step for the datasets. The preprocessing stage includes the following steps: separating hyperlinks from the adjacent text, normalize twitter-specific tokens, extracting text from '*' (e.g. *good* − > good), replacing & symbol, lower-casing the text, normalize multiple occurrences of vowels and consonants, normalize emojis and numbers, spiting 'number' and 'emoji' when adjacent to text, removing non-alphanumeric characters, removing very long words $>= 15$ and short words < 2 to reduce sparsity, removing multiple sequential occurrences of the same token.

3.2 Under-sampling

As one of the main focus of this paper is to model imbalanced dataset, we have utilized random re-sampling (Napierała et al., 2010; Johnson and Khoshgoftaar, 2019) to balance the class distribution. After Tweet preprocessing, we have used random under-sampling of the majority class. Oversampling is another approach to balance the training dataset, however, often oversampling minority class instances does not fully balance the training data whereas under-sampling of majority class is found to be better at balancing the training data (Jamil, 2017). We have left experiments involving oversampling for future works.

For extremely imbalanced class distribution in task 1, the under-sampling was done in two steps. First, we have utilized an informed under-sampling technique based on a pre-trained named-entity recognizer (NER) (Andriy Mulyar and McInnes, 2018) trained on clinical notes[2]. The NER model was used to extract and remove the Tweets that have medical named entities represented by informal expressions (e.g. 'lol').

After that, random under-sampling of the majority class was done on the remaining Tweets. The balanced dataset size is twice the number of the positive label in the dataset. For example, task 1 has a total of 146 positive examples (i.e. minority class) and after applying the random under-sampling, the negative examples (i.e. majority class) reduced down to 146. The resulting balanced training dataset size is 292 (i.e. 146 + 146). This part of random down-sampling was applied for all the datasets with IR $< 50\%$. A python toolbox, imbalanced-learn (Lemaître et al., 2017), was used to re-balance the class distribution.

3.3 Sentence Embedding

Sentence embedding methods attempt to encode a variable-length input sentence into a fixed-length vector. While preprocessing and normalization helps in better syntactic representation (Kaufmann and Kalita, 2010; Kong et al., 2014), sentence embeddings were used to improve the semantic representation. Moreover, sentence embeddings have been used in sentence classification to address the class imbalance problem (Madabushi et al., 2019). In recent years, several sentence embedding methods, exploration of semantic properties of resulted embeddings and impact of embedding on the downstream application have been proposed (Le and Mikolov, 2014; Kiros et al., 2015; Pagliardini et al., 2018; Schwenk and Douze, 2017; Arora et al., 2019; Zhu et al., 2018). Among them Sent2Vec (Pagliardini et al., 2018) demonstrated the robustness of generated general-purpose sentence embeddings when transferred to a wide range of prediction benchmarks. Sent2Vec is an unsupervised sentence embedding technique allowing composing sentence embeddings using word vectors along with n-gram embeddings.

There are three distinct language-specific categorizations of the Tweets in the shared task datasets. Among the participated tasks, four are based on English Tweets whereas two tasks are based on French and Russian Tweets. For representing the four English language Tweets, we have used 700-dimensional pre-trained[3] Sent2Vec model trained with English Tweets incorporating bi-gram embedding. For non-English Tweets, we have used 1024-dimensional pre-trained Language-Agnostic Sentence Representations (LASER) embeddings[4] (Schwenk and Douze, 2017). The pre-trained LASER model was based on 93 languages and does not need a specification of the input language. The sentence encoder also supports code-switching, i.e. the same sentences can contain words in several different languages. Overall,

---

[2]https://github.com/NLPatVCU/medaCy_model_clinical_notes
[3]https://github.com/epfml/sent2vec
[4]https://github.com/facebookresearch/LASER

Table 2: Evaluation Score on Validation, Test Datasets and Mean Score Based on all Submissions

Task	Validation			Test			Mean (all submissions)		
	F1	P	R	F1	P	R	F1	P	R
1	0.08	0.04	**0.97**	0.05	0.03	**0.90**	0.66	0.70	**0.69**
2 (EN)	0.39	0.25	**0.84**	0.32	0.19	**0.87**	0.46	0.42	**0.59**
2 (FR)	0.08	0.04	**0.75**	0.07	0.04	**0.60**	0.07	-	-
2 (RU)	0.35	0.22	**0.86**	0.35	0.22	**0.89**	0.43	0.36	**0.58**
3 (D)	-	-	-	0.159	**0.178**	0.143	0.564	**0.607**	0.557
4	0.45	0.36	**0.62**	0.46	0.35	**0.68**	0.49	-	-

Sent2Vec model was used to embed the English Tweets, whereas LASER was used for multilanguage tasks.

3.4   Classification Model Selection

We mainly focus on traditional systems to classify the Tweets. In recent years, deep learning models have shown superior performance in classification tasks (Weissenbacher et al., 2019). However, neural models require longer training time (Livni et al., 2014) and due to time constraint, we have utilized traditional baselines from related works (Weissenbacher et al., 2019). The multi-class classification is modeled as an one-vs-one scheme. We have applied Support Vector Machine (SVM) (Cortes and Vapnik, 1995) with radial basis function kernel and tree-based ensemble models, such as, Extra Trees Classifier (Geurts et al., 2006), Random Forest Classifier (Breiman, 2001) for data modeling. The features for models were learned by sentence embedding models as described in Sec. 3.3 and the final model was chosen based on the 10 split k-fold cross-validation on the down-sampled training dataset. For most of the tasks, the SVM model provided better result based on 10-fold cross-validation with mean F1-score evaluation metric, except for T2 (Fr) task where Extra Tree Classifier gave best mean F-1 score. Based on the experiments, for evaluation of the shared tasks, Extra Tree Classifier was used for T2 (Fr), and SVM models were used for all other participated tasks.

4   Methodology for Span Detection Task

Task 3 can be divided into two sub-tasks: span detection and concept normalization. We have worked on the detection part of the task and the normalization part is left for the future work. The dictionary-based simple traditional approach has been used in information detection task (Egorov et al., 2004). The approach utilizes a carefully constructed dictionary to identify and tag the related entities from the Tweets. For task 3, we have used a dictionary-based search approach to detect the span of adverse effects. A dictionary was generated from the AMIA shared task[5] with 6,649 annotated instances[6] of adverse effects.

5   Results and Analysis

The official performance evaluation metrics for the tasks are precision (P), recall (R), and F1-score (F1) computed on the positive class (i.e. minority class). Table 2 reports the performance scores of the tasks described in Tab. 1. The rows of the table correspond to the tasks and the columns correspond to the evaluation scores on validation and test datasets. The mean test scores of all the submissions are also reported in the table. The analysis of the results is based on the validation dataset.

Using the methodology described in this paper, from the table we can observe that the recall score is higher compared to the precision score in most cases. As positive labels refer to the minority class, there are a larger number of negative examples that could become false positives (FP). Conversely, there are fewer positive examples that could become false negatives (FN). The disproportion in class distribution

---

can result in the disproportion of the FPs and FNs which could result in high recall and low precision score. As F1-score is a weighted combination of these two matrices, a low precision score can cause an overall low F1-score. Although random under-sampling was used to balance the training dataset, it does not improve the precision score on the validation and test dataset since the validation and test dataset are still imbalanced that reflect the real class distribution.

6 Conclusion

This paper reports the preliminary exploration of Twitter-based imbalanced data modeling techniques for health applications. We have explored techniques for efficient syntactic and semantic representation of Tweets. As the focus was on imbalanced class distribution, random under-sampling with novel noise reduction technique was utilized to balance the dataset. The traditional margin-based and ensemble tree-based classifiers were used to classify the Tweets.

The directions for future works involve deeper exploration and extensive methodological improvements to enhance the learning performance. For example, the statistical learning theorems establish the number of examples needed to estimate the accuracy of a classifier as a function of its complexity (VC-dimension). However, the class imbalance does not enter these formulas anywhere (Juba and Le, 2019). Overall, a detailed exploration of classification models and evaluation metrics can be done to enhance model performance for skewed data distribution.

Acknowledgements

The author would like to acknowledge the support of the Business and Local Government Data Research Centre (ES/S007156/1) funded by the Economic and Social Research Council (ESRC) for undertaking this work.

References

Wasim Ahmed, Peter A Bath, and Gianluca Demartini. 2017. Using twitter as a data source: An overview of ethical, legal, and methodological challenges. In *The Ethics of Online Research*. Emerald Publishing Limited.

Natassja Lewinski Andriy Mulyar and Bridget McInnes. 2018. Tac srie 2018: Extracting systematic review information with medacy. *National Institute of Standards and Technology (NIST) 2018 Systematic Review Information Extraction (SRIE) ¿ Text Analysis Conference*, nov.

Sanjeev Arora, Yingyu Liang, and Tengyu Ma. 2019. A simple but tough-to-beat baseline for sentence embeddings. In *5th International Conference on Learning Representations, ICLR 2017*.

Alexandra Balahur. 2013. Sentiment analysis in social media texts. In *Proceedings of the 4th workshop on computational approaches to subjectivity, sentiment and social media analysis*, pages 120–128.

Gustavo EAPA Batista, Ronaldo C Prati, and Maria Carolina Monard. 2004. A study of the behavior of several methods for balancing machine learning training data. *ACM SIGKDD explorations newsletter*, 6(1):20–29.

Russell Beckley. 2015. Bekli: A simple approach to twitter text normalization. In *Proceedings of the Workshop on Noisy User-generated Text*, pages 82–86.

Leo Breiman. 2001. Random forests. *Machine Learning*, 45(1):5–32.

Mateusz Buda, Atsuto Maki, and Maciej A Mazurowski. 2018. A systematic study of the class imbalance problem in convolutional neural networks. *Neural Networks*, 106:249–259.

Corinna Cortes and Vladimir Vapnik. 1995. Support-vector networks. *Machine learning*, 20(3):273–297.

Yin Cui, Menglin Jia, Tsung-Yi Lin, Yang Song, and Serge Belongie. 2019. Class-balanced loss based on effective number of samples. In *Proceedings of the IEEE Conference on Computer Vision and Pattern Recognition*, pages 9268–9277.

Thierry Declerck and Piroska Lendvai. 2015. Processing and normalizing hashtags. In *Proceedings of the International Conference Recent Advances in Natural Language Processing*, pages 104–109, Hissar, Bulgaria, September. INCOMA Ltd. Shoumen, BULGARIA.

Daantje Derks, Arjan ER Bos, and Jasper Von Grumbkow. 2008. Emoticons and online message interpretation. *Social Science Computer Review*, 26(3):379–388.

Sergei Egorov, Anton Yuryev, and Nikolai Daraselia. 2004. A simple and practical dictionary-based approach for identification of proteins in medline abstracts. *Journal of the American Medical Informatics Association*, 11(3):174–178.

Pierre Geurts, Damien Ernst, and Louis Wehenkel. 2006. Extremely randomized trees. *Machine learning*, 63(1):3–42.

Guo Haixiang, Li Yijing, Jennifer Shang, Gu Mingyun, Huang Yuanyue, and Gong Bing. 2017. Learning from class-imbalanced data: Review of methods and applications. *Expert Systems with Applications*, 73:220–239.

Haibo He and Edwardo A Garcia. 2009. Learning from imbalanced data. *IEEE Transactions on knowledge and data engineering*, 21(9):1263–1284.

Zunaira Jamil. 2017. *Monitoring tweets for depression to detect at-risk users*. Ph.D. thesis, Université d'Ottawa/University of Ottawa.

Justin M Johnson and Taghi M Khoshgoftaar. 2019. Survey on deep learning with class imbalance. *Journal of Big Data*, 6(1):27.

Brendan Juba and Hai S Le. 2019. Precision-recall versus accuracy and the role of large data sets. In *Proceedings of the AAAI Conference on Artificial Intelligence*, volume 33, pages 4039–4048.

Max Kaufmann and Jugal Kalita. 2010. Syntactic normalization of twitter messages. In *International conference on natural language processing, Kharagpur, India*, volume 16.

Ryan Kiros, Yukun Zhu, Russ R Salakhutdinov, Richard Zemel, Raquel Urtasun, Antonio Torralba, and Sanja Fidler. 2015. Skip-thought vectors. In *Advances in neural information processing systems*, pages 3294–3302.

Ari Z. Klein, Ilseyar Alimova, Ivan Flores, Arjun Magge, Zulfat Miftahutdinov, Anne-Lyse Minard, Karen O'Connor, Abeed Sarker, Elena Tutubalina, Davy Weissenbacher, and Graciela Gonzalez-Hernandez. 2020. Overview of the fifth social media mining for health applications (# smm4h) shared tasks at coling 2020. *In Proceedings of the Fifth Social Media Mining for Health Applications (# SMM4H) Workshop & Shared Task*.

Lingpeng Kong, Nathan Schneider, Swabha Swayamdipta, Archna Bhatia, Chris Dyer, and Noah A Smith. 2014. A dependency parser for tweets. In *Proceedings of the 2014 Conference on Empirical Methods in Natural Language Processing (EMNLP)*, pages 1001–1012.

Guillaume Lample, Miguel Ballesteros, Sandeep Subramanian, Kazuya Kawakami, and Chris Dyer. 2016. Neural architectures for named entity recognition. In *Proceedings of the 2016 Conference of the North American Chapter of the Association for Computational Linguistics: Human Language Technologies*, pages 260–270.

Quoc Le and Tomas Mikolov. 2014. Distributed representations of sentences and documents. In Eric P. Xing and Tony Jebara, editors, *Proceedings of the 31st International Conference on Machine Learning*, volume 32 of *Proceedings of Machine Learning Research*, pages 1188–1196, Bejing, China, 22–24 Jun. PMLR.

Guillaume Lemaître, Fernando Nogueira, and Christos K. Aridas. 2017. Imbalanced-learn: A python toolbox to tackle the curse of imbalanced datasets in machine learning. *Journal of Machine Learning Research*, 18(17):1–5.

Charles X. Ling and Victor S. Sheng, 2010. *Class Imbalance Problem*, pages 171–171. Springer US, Boston, MA.

Roi Livni, Shai Shalev-Shwartz, and Ohad Shamir. 2014. On the computational efficiency of training neural networks. In *Advances in neural information processing systems*, pages 855–863.

Harish Tayyar Madabushi, Elena Kochkina, and Michael Castelle. 2019. Cost-sensitive bert for generalisable sentence classification with imbalanced data. *EMNLP-IJCNLP 2019*, page 125.

Éric Marchand and William E Strawderman. 2020. On shrinkage estimation for balanced loss functions. *Journal of Multivariate Analysis*, 175:104558.

David Nadeau and Satoshi Sekine. 2007. A survey of named entity recognition and classification. *Linguisticae Investigationes*, 30(1):3–26, January. Publisher: John Benjamins Publishing Company.

Krystyna Napierała, Jerzy Stefanowski, and Szymon Wilk. 2010. Learning from imbalanced data in presence of noisy and borderline examples. In *International Conference on Rough Sets and Current Trends in Computing*, pages 158–167. Springer.

Matteo Pagliardini, Prakhar Gupta, and Martin Jaggi. 2018. Unsupervised Learning of Sentence Embeddings using Compositional n-Gram Features. In *NAACL 2018 - Conference of the North American Chapter of the Association for Computational Linguistics*.

M Mostafizur Rahman and Darryl N Davis. 2013. Addressing the class imbalance problem in medical datasets. *International Journal of Machine Learning and Computing*, 3(2):224.

Rupsa Saha, Abir Naskar, Tirthankar Dasgupta, and Lipika Dey. 2018. Leveraging web based evidence gathering for drug information identification from tweets. In *Proceedings of the 2018 EMNLP Workshop SMM4H: The 3rd Social Media Mining for Health Applications Workshop & Shared Task*, pages 67–69.

Abeed Sarker, Maksim Belousov, Jasper Friedrichs, Kai Hakala, Svetlana Kiritchenko, Farrokh Mehryary, Sifei Han, Tung Tran, Anthony Rios, Ramakanth Kavuluru, Berry de Bruijn, Filip Ginter, Debanjan Mahata, Saif M Mohammad, Goran Nenadic, and Graciela Gonzalez-Hernandez. 2018. Data and systems for medication-related text classification and concept normalization from Twitter: insights from the Social Media Mining for Health (SMM4H)-2017 shared task. *Journal of the American Medical Informatics Association*, 25(10):1274–1283, 10.

Holger Schwenk and Matthijs Douze. 2017. Learning joint multilingual sentence representations with neural machine translation. In *Proceedings of the 2nd Workshop on Representation Learning for NLP*, pages 157–167.

Luchen Tan, Haotian Zhang, Charles Clarke, and Mark Smucker. 2015. Lexical comparison between wikipedia and twitter corpora by using word embeddings. In *Proceedings of the 53rd Annual Meeting of the Association for Computational Linguistics and the 7th International Joint Conference on Natural Language Processing (Volume 2: Short Papers)*, pages 657–661.

Gary M Weiss. 2004. Mining with rarity: a unifying framework. *ACM Sigkdd Explorations Newsletter*, 6(1):7–19.

Davy Weissenbacher, Abeed Sarker, Arjun Magge, Ashlynn Daughton, Karen OConnor, Michael Paul, and Graciela Gonzalez. 2019. Overview of the fourth social media mining for health (smm4h) shared tasks at acl 2019. In *Proceedings of the Fourth Social Media Mining for Health Applications (# SMM4H) Workshop & Shared Task*, pages 21–30.

Tom Young, Devamanyu Hazarika, Soujanya Poria, and Erik Cambria. 2018. Recent trends in deep learning based natural language processing. *IEEE Computational intelligenCe magazine*, 13(3):55–75.

Xunjie Zhu, Tingfeng Li, and Gerard De Melo. 2018. Exploring semantic properties of sentence embeddings. In *Proceedings of the 56th Annual Meeting of the Association for Computational Linguistics (Volume 2: Short Papers)*, pages 632–637.

HITSZ-ICRC: A Report for SMM4H Shared Task 2020-Automatic Classification of Medications and Adverse Effect in Tweets

Xiaoyu Zhao, Ying Xiong, Buzhou Tang

Department of Computer Science, Harbin Institute of Technology (Shenzhen), China

`zhaoxiaoyux@126.com, {xiongying0929, tangbuzhou}@gmail.com`

Abstract

This is the system description of the Harbin Institute of Technology Shenzhen (HITSZ) team for the first and second subtasks of the fifth Social Media Mining for Health Applications (SMM4H) shared task in 2020. The first task is automatic classification of tweets that mention medications and the second task is automatic classification of tweets in English that report adverse effects. The system we propose for these tasks is based on bidirectional encoder representations from transformers (BERT) incorporating with knowledge graph and retrieving evidence from online information. Our system achieves an F1 of 0.7553 in task 1 and an F1 of 0.5455 in task 2.

1 Introduction

With greatly increasing number of social media users, plenty of personal information has been posted in social networking websites such as Twitter, including massive data about health and medicine. Social media has attracted more and more attention from researchers and industrials in the health and medicine domain. To utilize the imbalanced and informal expressions of clinical concepts, a system should be able to alleviate the noise of social media posts and detect medical mentions which requires professional knowledge. In 2020, the health language processing lab at University of Pennsylvania organized the Social Media Mining for Health Applications (SMM4H) shared task about using social media mining for health monitoring, which includes 5 subtasks. We participate in the first two subtasks: (1) Automatic classification of tweets that mention medications, and (2) Automatic classification of tweets in English that report adverse effects (AEs). In this report, we briefly introduce our system developed for the two subtasks. The system is based on pre-trained language model BERT (Devlin, Chang, Lee, & Toutanova, 2018) with external knowledge.

2 Data and Pre-processing

The 2020 SMM4H organizers provide a dataset of tweets with a label indicating whether the tweets mention medications for task1 (1-Yes, 0-No), and a dataset of tweets with a label indicating whether the tweets mention AEs for task2. Each tweet includes text, a date and an identified ID. Table 1 describes the statistics of the two individual datasets, where #1 and #0 denote the numbers of tweets that mention medications and not respectively, #all denotes the total number of tweets, and NA denotes that we do not know the number of those tweets.

Task	Dataset	#1	#0	#all
Task1	training set	181	69,091	69,272
	test set	NA	NA	29,687
Task2	training set	2,374	23,298	25,672
	test set	NA	NA	4,659

Table 1: Distribution of label over the training and test datasets of task 1 and task 2.

Proceedings of the 5th Social Media Mining for Health Applications (#SMM4H) Workshop & Shared Task, pages 146–149
Barcelona, Spain (Online), December 12, 2020.

For both tasks, we pre-process all tweets via removing the usernames, URLs, emoticons, non-ASCII characters and repeated punctuations.

3 System Description

Our system for both task 1 and task 2 is based on BERT, which is used to represent tweet. Below we describe in detail the methods for the two tasks one by one.

3.1 Methods for task1

In this task, we first deploy BERT directly (Shuai Chen, 2019), and then extend it to BERT_Med by introducing potential medications from DrugBank[1]. The architecture of BERT_Med is shown in Fig. 1, where FC is a fully-connected layer. Given a piece of tweet, we first calculate its similarities with medications, and then select those medications with similarity less than 0.1 as its next tweet, finally input the tweet and the selected medications into BERT. The similarity between tweets and medications is measured by cosine using TF-IDF(Term Frequency-Inverse Document Frequency) as their representations.

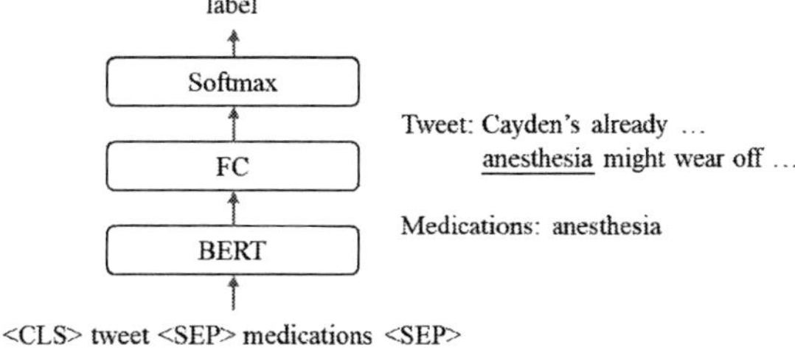

Figure 1: BERT with potential drugs (BERT_Med)

3.2 Methods for task2

For task 2, we propose the following two models based on BERT as shown in Fig. 2.

Strategy 1: To extract the internal information of AEs in tweets, we add the potential AEs corresponding to the drug mentions detected in the tweets. Drug mentions in tweets are detected by exact matching on Drugbank and the drug dictionary (Azadeh N, Abeed S, O'Connor Karen, et al. 2015), and the AEs come from MedlinePlus[2]. Similar to BERT_MED in Fig. 1, we treat the potential AEs as the next tweet of the given tweets and input them into BERT. In addition, we add a self-attention module (Ashish Vaswani, Noam Shazeer, Niki Parmar and et al. 2017) after BERT to redistribute representations of the tweet and potential AEs.

Strategy 2: To further integrate external knowledge into strategy 1, we introduce medication representation learnt from MeSH[3] by TransE (Bordes A, Usunier N, Garcia-Duran A, et al. 2013) as shown in Fig. 2. A similar self-attention module is employed to extract information from the concatenated entity embeddings learnt by TransE. Finally, we feed combination of the final representation of the texts and the medication entities into a linear transformation layer for the final prediction.

---

[1] https://www.drugbank.ca/
[2] https://medlineplus.gov/
[3] https://www.nlm.nih.gov/mesh/meshhome.html

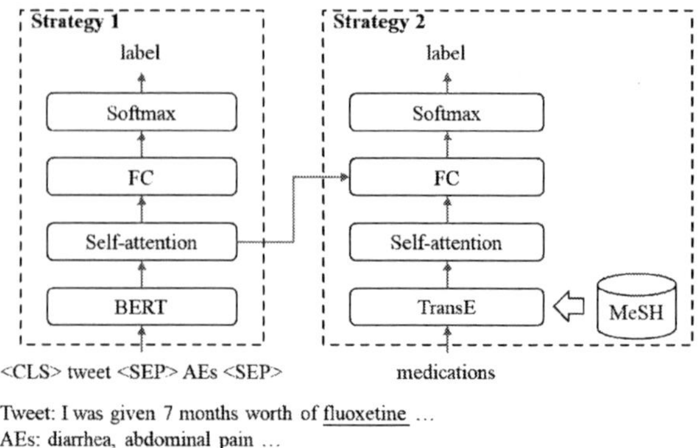

Figure 2: Model architectures in task 2.

4 Experiments

For task1, we set batch size to 64, learning rate to 2e-5 and sequence length to 64 when training models. For task2, the batch size and sequence length of BERT are the same as the ones of task1 while the learning rates of BERT and attention layer are set to 1e-5 and 5e-5 respectively. We train all models using 5-fold cross validation on the training set.

5 Results

Task	System	Training set			Test set		
		F1	P	R	F1	P	R
Task 1	BERT	0.8436	0.9310	0.7714	0.7553	0.8182	0.7013
	BERT_Med	0.6667	0.5246	0.9143	NA	NA	NA
Task 2	BERT_Strategy1	0.6877	0.6501	0.7299	0.5190	0.4393	0.6340
	BERT_Strategy2	0.6802	0.6823	0.6741	0.5254	0.4460	0.6392
	BERT_Merged	0.6701	0.6177	0.7321	0.5455	0.4701	0.6495

Table 2: Results on training and test sets for task 1 and task 2.

The results of our system provided by the organizers are listed in Table 2, where NA denotes that we do not submit results as internet problem, BERT_Merged is the model that merges results of BERT_Strategy1 and BERT_Strategy2 by voting. Our system achieves the highest F-score of 0.7553 on task1 and 0.5455 on task2. For task 1, on the training set, BERT_Med achieves much higher recall, but much lower precision than BERT. For task2, BERT_Strategy1 and BERT_Strategy2 perform roughly the same on the training and test sets. BERT_Merged performs slightly worse than them on the training set, but a little higher on the test set. The difference between performances of our system on the training set and test set is very large.

Reference

Jacob Devlin, Ming-Wei Chang, Kenton Lee, and Kristina Toutanova. 2018. *Bert: Pre-training of deep bidirectional transformers for language understanding.* arXiv preprint arXiv:1810.04805.

Shuai Chen, Yuanhang Huang, Xiaowei Huang, Haoming Qin, Jun Yan, Buzhou Tang. 2019. *HITSZ-ICRC: A report for SMM4H shared task 2019-automatic classification and extraction of adverse effect mentions in tweets.* Proceedings of the Fourth Social Media Mining for Health Applications (# SMM4H) Workshop & Shared Task. pp. 47-51.

Azadeh N, Abeed S, O'Connor Karen, et al. 2015. *Pharmacovigilance from social media: mining adverse drug reaction mentions using sequence labeling with word embedding cluster features.* Journal of the American Medical Informatics Association. (3):3.

Ashish Vaswani, Noam Shazeer, Niki Parmar, Jakob Uszkoreit, Llion Jones, Aidan N. Gomez, Lukasz Kaiser and Illia Polosukhin. 2017. Attention Is All You Need. arXiv preprint arXiv: 1706.03762.

Bordes A, Usunier N, Garcia-Duran A, et al. 2013. *Translating embeddings for modeling multi-relational data.* Proc of NIPS. Cambridge, MA: MIT Press, 2013:2787-2795

Automatic classification of tweets mentioning a medication using pre-trained sentence encoders

Laiba Mehnaz
MIDAS Lab, IIIT-Delhi
laibamehnazbt2k16@dtu.ac.in

Abstract

This paper describes our submission to the 5th edition of the Social Media Mining for Health Applications (SMM4H) shared task 1. Task 1 aims at the automatic classification of tweets that mention a medication or a dietary supplement. This task is specifically challenging due to its highly imbalanced dataset, with only 0.2% of the tweets mentioning a drug. For our submission, we particularly focused on several pretrained encoders for text classification. We achieve an F1 score of 0.75 for the positive class on the test set.

1 Introduction

Automatic drug name recognition has mostly been studied in terms of extracting drug names from medical documents and biomedical articles (Liu et al., 2015). However, expanding the same task to extracting drug names from tweets poses a lot more challenges. Tweets are shorter and do not provide enough context compared to academic biomedical articles; they also contain ambiguity, noise, and misspellings, especially in the form of colloquially used terms for the same drugs (Weissenbacher et al., 2019). The shared task of the 5th Social Media Mining for Health Applications specifically aims at tasks that use natural language processing for health applications. We participated in task 1, which is defined as the automatic classification of tweets that mention medications. We use several pre-trained encoders such as BERT (Devlin et al., 2019), BioBERT (Lee et al., 2019), Clinical BioBERT (Alsentzer et al., 2019), SciBERT (Beltagy et al., 2019), RoBERTa (Liu et al., 2019), BioMed-RoBERTa (Gururangan et al., 2020), ELECTRA (Clark et al., 2020) and ERNIE 2.0 (Sun et al., 2019) for the classification of tweets.

2 Dataset

The training and the validation dataset were provided to us by the organizers of SMM4H2020. The training dataset consisted of 55419 tweets, with only 146 positive tweets and 55273 negative tweets. The validation dataset consisted of 13853 tweets, with only 35 positive tweets and 13818 negative tweets. The dataset is highly imbalanced, and the positive tweets account for only 0.2% of the whole dataset. The test set for submitting our system predictions consisted of 29687 tweets.

3 Experiments and System Descriptions

Along with BERT, we use several other pre-trained sentence encoders that are a result of various improvisations over BERT such as BioBERT (Lee et al., 2019), Clinical BioBERT (Alsentzer et al., 2019), SciBERT (Beltagy et al., 2019), RoBERTa (Liu et al., 2019), and BioMed-RoBERTa (Gururangan et al., 2020). For all of the above pre-trained sentence encoders, we use the PyTorch implementation through the transformers library[1] fine-tuning them for 3 epochs with a learning rate of 2e-5, maximum sequence length as 128, and a batch size of 8. Unlike BERT, ELECTRA (Clark et al., 2020) uses an alternative pre-training task, called replaced token detection. It aims to be more sample-efficient than masked language

[1]https://github.com/huggingface/transformers

Proceedings of the 5th Social Media Mining for Health Applications (#SMM4H) Workshop & Shared Task, pages 150–152
Barcelona, Spain (Online), December 12, 2020.

modeling used in BERT. Using the implementation of ELECTRA[2] provided by the authors we fine-tuned ELECTRA for 3 epochs with a learning rate of 1e-4, maximum sequence length as 128, and a batch size of 32. ERNIE 2.0 (Sun et al., 2019) provides a continual pre-training framework to incrementally build several pre-training tasks that focus on extracting lexical, syntactic, and semantic information from the training corpora. We use the implementation in PaddlePaddle[3] provided by the authors and fine-tune ERNIE 2.0 for 3 epochs with a learning rate of 3e-5, maximum sequence length as 128, and a batch size of 64.

Model name	F1 score	Precision	Recall
BERT base	0.78	0.83	0.74
BioBERT base	0.80	0.84	0.77
Clinical BioBERT base	0.81	0.82	0.80
SciBERT base	0.83	0.90	0.77
RoBERTa base	0.81	0.82	0.80
BioMed-RoBERTa base	0.85	0.90	0.80
ELECTRA base	0.79	0.81	0.77
ERNIE 2.0	0.83	0.92	0.74

Table 1: F1 score, Precision, and Recall for the positive class on the validation dataset.

Model name	F1 score	Precision	Recall
BioMed-RoBERTa base	0.755	0.770	0.740

Table 2: F1 score, Precision, and Recall for the positive class on the test dataset.

4 Results and Discussion

Due to the imbalance in the dataset, the metric used for evaluating the systems is the F1 score for the positive class, where positive class refers to the set of tweets that mention a drug or a dietary supplement. Table 1 contains the scores of the pre-trained sentence encoders on the validation dataset. As can be seen in Table 1, BERT has the worst performance of all the models. And, BioMed-RoBERTa has the best performance. ELECTRA base performs slightly better than BERT. Compared to BERT, there is a consistent increase in performance of all the models that used domain-specific data for pre-training. Within the group of models using biomedical related data for pre-training, SciBERT performs better than both BioBERT and Clinical BioBERT. SciBERT is trained on a multi-domain corpus, where papers from the computer science domain account for 18% of all the papers, and papers from the biomedical domain account for 82%. Unlike BioBERT and Clinical BioBERT, SciBERT is trained from scratch and has its own vocabulary called the scivocab. These factors could be the reason for SciBERT's better performance compared to both BioBERT and Clinical BioBERT. BioMed-RoBERTa's performance is visibly better than SciBERT. This performance increase could be due to RoBERTa's superior performance over BERT, as well as the additional pre-training data consisting of 2.68M full-text papers from S2ORC (Lo et al., 2020). It is interesting to note that RoBERTa's performance is comparable to BioBERT and Clinical-BERT without any domain-specific pre-training. It is also worth noting that ERNIE 2.0 has the same F1 score as SciBERT without any domain-specific pre-training. It also performs better than RoBERTa. ERNIE 2.0 seems to have learned better representations without any domain-specific training, which could be due to its variety of pre-training tasks aiming to capture lexical, syntactic, and semantic information from the dataset. ERNIE 2.0's performance also leads to an interesting question of the possibility of universal pre-trained models. Table 2 shows the results of our system prediction on the test set. Due

---

[2] https://github.com/google-research/electra

[3] https://github.com/PaddlePaddle/ERNIE

to time constraints, we could submit only one system prediction for the test dataset. Looking at the performance on the validation dataset, we chose to submit the predictions of BioMed-RoBERTa, as it gave the best performance on the validation dataset. Our system predictions on the test set are competitive and achieve above-average scores among the participants' systems.

References

Davy Weissenbacher, Abeed Sarker, Ari Klein, Karen O'Connor, Arjun Magge and Graciela Gonzalez-Hernandez. Deep neural networks ensemble for detecting medication mentions in tweets. *Journal of the American Medical Informatics Association*,Volume 26, Issue 12, December 2019, Pages 1618–1626, https://doi.org/10.1093/jamia/ocz156

Emily Alsentzer, John R. Murphy, Willie Boag, Wei-Hung Weng,Di Jin, Tristan Naumann and Matthew B. A. McDermott. (2019). Publicly Available Clinical BERT Embeddings. *ArXiv, abs/1904.03323*.

Iz Beltagy, Arman Cohan and Kyle Lo. (2019). SciBERT: Pretrained Contextualized Embeddings for Scientific Text. *ArXiv, abs/1903.10676*.

Jacob Devlin, Ming-Wei Chang, Kenton Lee and Kristina Toutanova. (2019). BERT: Pre-training of Deep Bidirectional Transformers for Language Understanding. *ArXiv, abs/1810.04805*.

Jinhyuk Lee, Wonjin Yoon, Sungdong Kim, Donghyeon Kim, Sunkyu Kim, Chan Ho So and Jaewoo Kang. (2020). BioBERT: a pre-trained biomedical language representation model for biomedical text mining. *Bioinformatics*.

Kevin Clark, Minh-Thang Luong, Quoc V. Le and Christopher D. Manning. (2020). ELECTRA: Pre-training Text Encoders as Discriminators Rather Than Generators. *ArXiv, abs/2003.10555*.

Kyle Lo, Lucy Lu Wang, Mark E Neumann, Rodney Michael Kinney and Daniel S. Weld. (2020). S2ORC: The Semantic Scholar Open Research Corpus. *ACL*.

Shengyu Liu, Buzhou Tang, Qingcai Chen and Xiaolong Wang. Drug Name Recognition: Approaches and Resources. *Information*. 2015; 6(4):790-810.

Suchin Gururangan, Ana Marasović, Swabha Swayamdipta,Kyle Lo, Iz Beltagy, Doug Downey and Noah A. Smith. (2020). Don't Stop Pretraining: Adapt Language Models to Domains and Tasks. *ACL*.

Yinhan Liu, Myle Ott, Naman Goyal, Jingfei Du, Mandar Joshi, Danqi Chen, Omer Levy, Mike Lewis, Luke Zettlemoyer and Veselin Stoyanov. (2019). RoBERTa: A Robustly Optimized BERT Pretraining Approach. *ArXiv, abs/1907.11692*.

Yu Sun, Shuohuan Wang, Yukun Li, Shikun Feng, Hao Tian, Hua Wu and Haifeng Wang. (2020). ERNIE 2.0: A Continual Pre-training Framework for Language Understanding. *AAAI*.

Approaching SMM4H 2020 with Ensembles of BERT Flavours

George-Andrei Dima[1,2], Andrei-Marius Avram[1,3],
Dumitru-Clementin Cercel[1]
University Politehnica of Bucharest[1]
Military Technical Academy Ferdinand I[2]
Research Institute for Artificial Intelligence, Romanian Academy[3]
andrei.dima@mta.ro, avram.andreimarius@gmail.com,
dumitru.cercel@upb.ro

Abstract

This paper describes our solutions submitted to the Social Media Mining for Health Applications (#SMM4H) Shared Task 2020. We participated in the following tasks: Task 1 aimed at classifying if a tweet reports medications or not, Task 2 (only for the English dataset) aimed at discriminating if a tweet mentions adverse effects or not, and Task 5 aimed at recognizing if a tweet mentions birth defects or not. Our work focused on studying different neural network architectures based on various flavors of bidirectional Transformers (i.e., BERT), in the context of the previously mentioned classification tasks. For Task 1, we achieved an F1-score (70.5%) above the mean performance of the best scores made by all teams, whereas for Task 2, we obtained an F1-score of 37%. Also, we achieved a micro-averaged F1-score of 62% for Task 5.

1 Introduction

In recent years, researchers around the world came to realize the usefulness of social media data for extracting health information. The Social Media Mining for Health Applications (#SMM4H) Shared Task 2020 (Klein et al., 2020) brings to the forefront the problem of extracting information from health social media posts. SMM4H was also organized in the previous years and involved several tasks, such as automatic detection of tweets mentioning medication, of tweets describing medication intake or adverse reactions, of tweets mentioning vaccination behaviour, as well as tasks on extraction and normalization of adverse effects utterances (Weissenbacher et al., 2018; Weissenbacher et al., 2019b). This year's competition had five tasks and our team focused on Tasks 1, 2, and 5.

Task 1 was a binary classification one that involved distinguishing between tweets in which some medications or dietary supplements were mentioned (the positive class) and other tweets (the negative class). The challenge of this task, unlike the 2018 similar task, consisted in the highly imbalanced data sets, that is, the tweets had a distribution of the two classes similar with the one encountered in practice. Therefore, the positive class counted for only 0.2% of the examples.

Classification of multilingual tweets that report adverse effects (Task 2) was also a binary classification task that was divided in multiple subtasks, each one concerning a different language. Our team submitted solutions only for the English subtask. For both Tasks 1 and 2, the evaluation metric was the F1-score for the positive class. On the other hand, classification of tweets reporting a birth defect pregnancy outcome (Task 5) was a multi-class classification task with the following classes: *defect*, *possible defect*, and *non-defect*. For this task, the evaluation metric was the micro-averaged F1-score of the first two classes.

We started to approach the competition by studying methods for data preprocessing and for overcoming the class unbalancing issue. Afterwards, we tested multiple language models for classification and we contributed by further pre-training the best language model on social media data. Moreover, we increased the robustness of our approach by constructing an ensemble which managed to obtain 70.5% F1-score, whereas the average of the best scores for the Task 1 was 66.28%.

This paper is further structured as follows. Section 2 presents previous works that helped us in developing our solutions. Section 3 describes the proposed models. In Section 4, we show and interpret the results of the systems. Finally, Section 5 summarizes the conclusions of our work.

Proceedings of the 5th Social Media Mining for Health Applications (#SMM4H) Workshop & Shared Task, pages 153–157
Barcelona, Spain (Online), December 12, 2020.

2   Related Work

As related tasks were also organized in previous years of SMM4H, a significant amount of work has been done in developing solutions to the proposed problems. Therefore, previous studies already established some directions for appropriate methods to preprocess social media data, for practices to address imbalanced data sets, or for language models that are more effective for the given tasks. Thus, Ellendorff et al. (2019) showed which data preprocessing steps are more likely to obtain better results on tweets.

The challenge of learning from imbalanced data sets has been reviewed by Chawla et al. (2004). Their paper analyzed some general solutions like over-sampling and under-sampling and also offered useful guidelines in applying these techniques. Moreover, Khosla (2018) approached the problem by assigning different weights to the imbalanced classes and this method proved particularly useful in our experiments.

Mahata et al. (2019) used transfer learning approaches, showing that Bidirectional Encoder Representations from Transforms (BERT) (Devlin et al., 2018) and Universal Language Model Fine-Tuning (ULMFiT) (Howard and Ruder, 2018) are able to handle classification tasks in the medical domain. Also, Gondane (2019) leveraged the focus on biomedical language of BERT for Biomedical Text Mining (BioBERT) (Lee et al., 2020).

The issue of detecting medication mentions in social media data has been previously approached with impressive results. Weissenbacher et al. (2019a) showed that ensemble classifiers can achieve performance close to humans in recognizing mentions of medications in tweets, on balanced data sets. Wu et al. (2018) obtained notable results on social media data mentioning drugs using a neural network based on a multi-head self-attention mechanism.

3   Method

3.1   Text Preprocessing

Data gathered from social media (tweets in our case) implies a specific informal language that is closer to the spoken English, rather than the texts on platforms like Wikipedia. This type of text is rich in grammatical errors, abbreviations (e.g., *"cuz"* instead of *"because"*) and words (e.g., *"lol"* or *"idk"*) that are encapsulating their own meaning and cannot be found in usual dictionaries.

Before feeding this kind of data to a language model, a preliminary step must be done, i.e. preprocessing. It can be noted that this step might strip useful information. Fortunately, previous work (Ellendorff et al., 2019) gave a direction of which preprocessing would provide the best results. For our methods, the best results were obtained by using the following preprocessing steps:

- Replace all URLs with *"url"*;

- Replace all usernames with *"user"*;

- Remove all non-ASCII characters;

- Remove all HTML character references;

- Replace multiple white spaces with one space.

We also experimented with other techniques for spell correction via the Ekphrasis library (Baziotis et al., 2017) but, for our models, the results were not significantly improved.

3.2   Experiments

After preprocessing, data was fed to BERT-based language models that are related to the medical field, namely: BioBERT, Clinical BERT (Alsentzer et al., 2019), BioFLAIR (Sharma and Daniel Jr, 2019), and BioELMO (Jin et al., 2019). Among these four models, the best results were achieved using BioBERT.

BERT was bidirectionally pre-trained using two tasks: Masked Language Modeling and Next Sentence Prediction on a large corpus composed of English Wikipedia and BookCorpus. Starting from the pre-trained BERT, BioBERT was further pre-trained on a biomedical corpus composed of PubMed abstracts and articles.

For Task 1, we performed a series of experiments. First, we fine-tuned the last three layers of BioBERT-Base on the balanced data set of SMM4H 2018 (Weissenbacher et al., 2018), using the *early stopping* technique. Afterwards, we further fine-tuned our model on the actual training set of Task 1. Because this data set was highly unbalanced (55,273 negative examples and only 146 positive examples), the model tended to classify all examples as negative. To overcome this challenge, we experimented with several techniques: over-sampling the positive class, under-sampling the negative class, the *focal loss* (Lin et al., 2017), and adding weights for each class when computing the *binary crossentropy* loss. For the *class weighting* technique, we computed the weights using the formula proposed by Khosla (2018). Our results show that this technique obtains the best performance. We will further refer to this system as **BioBERT-ClassWeights**.

As we mentioned earlier, the language used in tweets is rather different of the language used in the corpora that BioBERT was pre-trained on. Therefore, it seemed intuitive to further pre-train the obtained system on data from social media. Due to the lack of considerable resources, we confined on using the English tweets from the data sets provided within the SMM4H 2020 shared task for the tasks 1, 2, 3, and 5, in order to form a corpus, and we pre-trained BioBERT on it, using the script provided on GitHub[1]. We used this language model in the same system, which we described above, and obtained **BioBERT-PretrainTweets**. Even though this method improved the results, training multiple models resulted in distant scores, thus showing that the model cannot be considered as having sufficient robustness.

In order to improve the robustness of the solution, we constructed two ensembles of multiple classifiers. **Ensemble 1** was formed from three models that performed well on the validation set: BioBERT-Base pre-trained on tweets, Clinical BERT and BioBERT-Large. For **Ensemble 2**, the training and validation sets were combined, shuffled and then splitted in five equally sized folds. Five models of *BioBERT-PretrainTweets* were fine-tuned as for 5-fold Cross-Validation and were afterwards used to form an ensemble. Both ensembles are deciding by averaging the outputs of the composing models.

For Tasks 2 - English and 5, we used *BioBERT-PretrainTweets* fine-tuned for each task. We should mention that for Task 5, concerning multi-class classification, we switched the loss function to *categorical crossentropy*. Yet, because *class weighting* did not improve the results, we decided not to use it.

4   Results

In the practice phase, we submitted predictions on the validation sets for Tasks 1 and 2 - English. For the first task, *BioBERT-ClassWeights* achieved an F1-score of 67.6% with a precision of 69.6% and a recall of 65.7%, while *BioBERT-PretrainTweets* obtained an F1-score of 77.61% with a precision of 81.2% and a recall of 74.2%. For Task 2 - English, *BioBERT-PretrainTweets* achieved 53.87% F1-score on the validation set.

In the evaluation phase, we submitted three solutions for the first task: one prediction from *BioBERT-PretrainTweets* and one prediction from each ensemble described above. The prediction of the *Ensemble 2* scored above the mean scores for Task 1. We also submitted one solution for each Task 2 - English and Task 5. Tables 1, 2, and 3 show the reported scores for each submission, alongside with the averaged score of best submissions of all teams that participated. The precision and recall for the first two submissions on the Task 1 were not reported by the organizers.

Model	F1-score	Precision	Recall
BioBERT-PretrainTweets	55%	-	-
Ensemble 1	66%	-	-
Ensemble 2	**70.5%**	79.03%	63.64%
Mean score	66.28%	70.32%	69.48%

Table 1: Test results and the average of best submissions for Task 1.

---

[1]https://github.com/google-research/bert/blob/master/run_pretraining.py.

Model	F1-score	Precision	Recall
BioBERT-PretrainTweets	37%	26%	60%
Mean score	46%	42%	59%

Table 2: Test result and the average of best submissions for Task 2 - English.

Model	F1-score	Precision	Recall
BioBERT-PretrainTweets	62%	56%	69%
Mean score	65%	62%	68%

Table 3: Test result and the average of best submissions for Task 5.

For Task 1, the scores on the test set indicate that using only *BioBERT-PretrainTweets* is not enough. It performed below average for all tasks, even though the validation scores would have suggested otherwise. On the other hand, the scores of the ensembles improved the prediction significantly. The score of *Ensemble 2* was above the average of the best scores, showing that our best system is an ensemble of BioBERT language models each fine-tunned on both the validation and training sets in a 5-folds manner. Even though the training set is large enough, the positive class is so poorly represented that previously mentioned techniques, which usually address small data sets, significantly improved our system.

5   Conclusion

In this paper, we experimented with ensembles of bidirectional Transformers in the context of social media texts and we studied the value that pre-trained BERT flavours, like BioBERT or ClinicalBERT, bring in solving classification tasks in the medical domain.

We succeeded in obtaining a score above the average of the best scores using *Ensemble 2* on Task 1 and we showed that BERT-based classifiers can give acceptable results even with highly unbalanced data sets. We also showed that a BERT-based language model, pre-trained for a rather colloquial language, improved the results on the given tasks of social media data. Our results on Task 1 show that, in cases where one of the classes is poorly represented, ensembles increase the prediction performance.

Further experiments should consider including an enlargement of the corpus of tweets used for pre-training BioBERT, so that the model will be more capable of representing this type of data. The next step would be to design a new BERT flavour pretrained on social media texts. Another future direction in addressing class imbalance might consist in using data augmentation in order to generate examples of the less represented class (Croce et al., 2020).

References

Emily Alsentzer, John R Murphy, Willie Boag, Wei-Hung Weng, Di Jin, Tristan Naumann, and Matthew McDermott. 2019. Publicly available clinical bert embeddings. *arXiv preprint arXiv:1904.03323*.

Christos Baziotis, Nikos Pelekis, and Christos Doulkeridis. 2017. Datastories at semeval-2017 task 4: Deep lstm with attention for message-level and topic-based sentiment analysis. In *Proceedings of the 11th International Workshop on Semantic Evaluation (SemEval-2017)*, pages 747–754, Vancouver, Canada, August. Association for Computational Linguistics.

Nitesh V Chawla, Nathalie Japkowicz, and Aleksander Kotcz. 2004. Special issue on learning from imbalanced data sets. *ACM SIGKDD explorations newsletter*, 6(1):1–6.

Danilo Croce, Giuseppe Castellucci, and Roberto Basili. 2020. Gan-bert: Generative adversarial learning for robust text classification with a bunch of labeled examples. In *Proceedings of the 58th Annual Meeting of the Association for Computational Linguistics*, pages 2114–2119.

Jacob Devlin, Ming-Wei Chang, Kenton Lee, and Kristina Toutanova. 2018. Bert: Pre-training of deep bidirectional transformers for language understanding. *arXiv preprint arXiv:1810.04805*.

Tilia Ellendorff, Lenz Furrer, Nicola Colic, Noëmi Aepli, and Fabio Rinaldi. 2019. Approaching smm4h with merged models and multi-task learning. In *Proceedings of the Fourth Social Media Mining for Health Applications (# SMM4H) Workshop & Shared Task*, pages 58–61. University of Zurich.

Shubham Gondane. 2019. Neural network to identify personal health experience mention in tweets using biobert embeddings. In *Proceedings of the Fourth Social Media Mining for Health Applications (# SMM4H) Workshop & Shared Task*, pages 110–113.

Jeremy Howard and Sebastian Ruder. 2018. Universal language model fine-tuning for text classification. *arXiv preprint arXiv:1801.06146*.

Qiao Jin, Bhuwan Dhingra, William Cohen, and Xinghua Lu. 2019. Probing biomedical embeddings from language models. In *Proceedings of the 3rd Workshop on Evaluating Vector Space Representations for NLP*, pages 82–89.

Sopan Khosla. 2018. Emotionx-ar: Cnn-dcnn autoencoder based emotion classifier. In *Proceedings of the Sixth International Workshop on Natural Language Processing for Social Media*, pages 37–44.

Ari Z. Klein, Ilseyar Alimova, Ivan Flores, Arjun Magge, Zulfat Miftahutdinov, Anne-Lyse Minard, Karen O'Connor, Abeed Sarker, Elena Tutubalina, Davy Weissenbacher, and Graciela Gonzalez-Hernandez. 2020. Overview of the fifth social media mining for health applications (#smm4h) shared tasks at coling 2020. In *Proceedings of the Fifth Social Media Mining for Health Applications (#SMM4H) Workshop Shared Task*.

Jinhyuk Lee, Wonjin Yoon, Sungdong Kim, Donghyeon Kim, Sunkyu Kim, Chan Ho So, and Jaewoo Kang. 2020. Biobert: a pre-trained biomedical language representation model for biomedical text mining. *Bioinformatics*, 36(4):1234–1240.

Tsung-Yi Lin, Priya Goyal, Ross Girshick, Kaiming He, and Piotr Dollár. 2017. Focal loss for dense object detection. In *Proceedings of the IEEE international conference on computer vision*, pages 2980–2988.

Debanjan Mahata, Sarthak Anand, Haimin Zhang, Simra Shahid, Laiba Mehnaz, Yaman Kumar, and Rajiv Shah. 2019. Midas@ smm4h-2019: Identifying adverse drug reactions and personal health experience mentions from twitter. In *Proceedings of the Fourth Social Media Mining for Health Applications (# SMM4H) Workshop & Shared Task*, pages 127–132.

Shreyas Sharma and Ron Daniel Jr. 2019. Bioflair: Pretrained pooled contextualized embeddings for biomedical sequence labeling tasks. *arXiv preprint arXiv:1908.05760*.

Davy Weissenbacher, Abeed Sarker, Michael Paul, and Graciela Gonzalez. 2018. Overview of the third social media mining for health (smm4h) shared tasks at emnlp 2018. In *Proceedings of the 2018 EMNLP Workshop SMM4H: The 3rd Social Media Mining for Health Applications Workshop & Shared Task*, pages 13–16.

Davy Weissenbacher, Abeed Sarker, Ari Klein, Karen O'Connor, Arjun Magge, and Graciela Gonzalez-Hernandez. 2019a. Deep neural networks ensemble for detecting medication mentions in tweets. *Journal of the American Medical Informatics Association*, 26(12):1618–1626.

Davy Weissenbacher, Abeed Sarker, Arjun Magge, Ashlynn Daughton, Karen O'Connor, Michael Paul, and Graciela Gonzalez. 2019b. Overview of the fourth social media mining for health (smm4h) shared tasks at acl 2019. In *Proceedings of the Fourth Social Media Mining for Health Applications (# SMM4H) Workshop & Shared Task*, pages 21–30.

Chuhan Wu, Fangzhao Wu, Junxin Liu, Sixing Wu, Yongfeng Huang, and Xing Xie. 2018. Detecting tweets mentioning drug name and adverse drug reaction with hierarchical tweet representation and multi-head self-attention. In *Proceedings of the 2018 EMNLP Workshop SMM4H: The 3rd Social Media Mining for Health Applications Workshop & Shared Task*, pages 34–37.

NLP@VCU: Identifying adverse effects in English tweets for unbalanced data

Darshini Mahendran, Cora Lewis and Bridget T. McInnes
Department of Computer Science
Virginia Commonwealth University
mahendrand@vcu.edu corammlewis@gmail btmcinnes@vcu.edu

Abstract

This paper describes our participation in the Social Media Mining for Health Application (SMM4H 2020) Challenge Track 2 for identifying tweets containing Adverse Effects (AEs). Our system uses Convolutional Neural Networks. We explore downsampling, oversampling, and adjusting the class weights to account for the imbalanced nature of the dataset. Our results showed downsampling outperformed oversampling and adjusting the class weights on the test set however all three obtained similar results on the development set.

1 Introduction

This paper describes our participation in Task 2 of the Social Media Mining for Health Application (SMM4H) 2020 challenge to automatically identify Adverse Effects (AE) in English tweets. To address this challenge, we explored a supervised binary classification system to automatically identify the AEs using Convolutional Neural Networks (CNNs). In order to deal with the unbalanced nature of the data set, we explored downsampling, oversampling, and utilization of the class weights.

2 Methods

In this section, we discuss our AE identification system. Our system can be found here[1].

Feature Representation. We pre-process the data as follows: 1) the HTML symbol *&* are removed as per (Úbeda et al., 2019); 2) double-quotes are removed; 3) hashtags, links, and usernames are replaced with the string "hashtag", "link" and "username" respectively as per (Cortes-Tejada et al., 2019); 4) emojis are substituted with a phrase that represents that emoji as per (Vydiswaran et al., 2019); 5) tweets are lowercased. Each word in the tweet is represented as an embedding. We evaluated three embedding types, GloVe (Pennington et al., 2014), word2vec (Mikolov et al., 2013) and FastText (Godin, 2019), trained over different corpora types. Our resulting system described here uses GloVe trained on Twitter.

Algorithm. We evaluated using Convolutional Neural Network (CNNs) as per (Úbeda et al., 2019). One of the beneficial properties of CNN is that it preserves the spatial orientation, in this case, the sequence of the words in the tweet. CNNs consist of four layers: 1) embedding layer - to encode words in sentences by real-valued vectors; 2) convolution layer - to get local features from each part of the input; 3) pooling layer - to extract the most relevant features, and 4) feed-forward layer - a fully connected layer to perform classification. For this, we first feed each tweet into CNN to learn the AE representation of the tweet. Second, we apply the convolution layer to learn the local features from the embedding vectors obtained from each word of a tweet. Next, we apply the max-pooling layer to extract the most important feature. Next, we unstack the volume into a flat vector and feed it into the fully connected feedforward layer. Finally, the fixed-length vector is fed into a softmax layer to perform the classification. For training, the classification error is back-propagated and the model is re-trained until the error is minimized. The weights of the matrix and bias are the parameters that get tuned until the optimized model is obtained.

[1]https://github.com/NLPatVCU/SMM4H

Proceedings of the 5th Social Media Mining for Health Applications (#SMM4H) Workshop & Shared Task, pages 158–160
Barcelona, Spain (Online), December 12, 2020.

Imbalance data. Due to the imbalanced nature of the dataset, we evaluated our algorithm using three methods to reduce the imbalance: downsampling, oversampling, and adjusting Keras class weights:

Run 1: To downsample, non-AE tweets are removed in order to reach an input ratio of AE tweets to non-AE tweets. Through experimentation, we discovered that the best downsampling ratio is 1 AE tweet for every 4 non-AE tweets.

Run 2: To oversample, the AE tweets are repeated a specific number of times. For example, if the data set is oversampled 3 times, there would be 3 copies of each AE tweet. Through experimentation, we learned that oversampling 6 times is the most effective number of oversamples.

Run 3: We used the class weights option in Keras which incorporates the ratio of how much an AE tweet should be valued compared to a non-AE tweet.(Chollet and others, 2015) For example, if class weight is 1 for the non-AE tweets and 20 for the AE tweets, Keras would treat each AE tweet like its worth twenty non-AE tweets. Through experimentation, we found that a class weight of 1 for non-AE tweets and a class weight of 10 for AE tweets is most effective.

3   Data

The training set contains 20,544 tweets: 1,903 tweets contain AEs and 18,641 tweets do not contain AEs. The development set contains 5,134 tweets: 4,660 tweets are negative and 474 are positive. The training and development data sets are both highly imbalanced containing a 1:10 ratio of AE tweets to non-AE tweets. The test set contains 4,759 tweets.

4   Results

In this section, we report the Precision (P), Recall (R), and F-measure (F) of our system on the SSM4H Task 2 data set for the three runs described above. Table 1 shows the results obtained evaluated over the development data (Development Results), and the reported results evaluated over the test data (Reported Results). The results were achieved using Train-Test. The precision and recall were returned only for the best performing run. The results on the development set showed that using downsampling (Run1) or oversampling (Run 2) obtained a higher precision and lower recall whereas using class weights (Run 3) obtained a higher recall and lower precision. The F-1 scores achieved with the development data were all very similar, with downsampling and class weights achieving identical F-1 scores. This was unlike in the test data results where downsampling (Run 1) showed a higher recall than precision and obtained the highest F-1 score of all the runs.

Run		Development Results			Reported Results		
	Dimension	Precision	Recall	F1 Score	Precision	Recall	F1 Score
Run 1 (downsample 1:4)	100	0.59	0.43	0.50	0.28	0.36	0.35
Run 2 (oversample 6)	50	0.56	0.44	0.49	-	-	0.34
Run 3 (class weights 10:1)	50	0.43	0.58	0.50	-	-	0.31

Table 1: Development and Evaluation Results

5   Conclusions and Future Work

Our CNN model achieved a reported F1 score of 0.35 on the test dataset and 0.50 on the development data. Our experimentation with pre-trained word embedding showed that GloVe trained on Twitter is optimal for this task. Our results also show that downsampling, oversampling, and Keras class weights achieve similar F1 scores with the development data, though downsampling outperformed oversampling and class weights on the test data. In the future, we would like to experiment further in the pre-processing stage. We also plan to explore the possibility of character embeddings or utilizing Recurrent Neural Networks (RNNs) in depth. We would like to investigate the usage of additional word embeddings such as Bidirectional Encoder Representations from Transformers (BERT) for this task as well.

References

François Chollet et al. 2015. Keras. https://keras.io.

Javier Cortes-Tejada, Juan Martinez-Romo, and Lourdes Araujo. 2019. Nlp@ uned at smm4h 2019: Neural networks applied to automatic classifications of adverse effects mentions in tweets. In *Proceedings of the Fourth Social Media Mining for Health Applications (# SMM4H) Workshop & Shared Task*, pages 93–95.

Fréderic Godin. 2019. *Improving and Interpreting Neural Networks for Word-Level Prediction Tasks in Natural Language Processing*. Ph.D. thesis, Ghent University, Belgium.

Tomas Mikolov, Ilya Sutskever, Kai Chen, Greg S Corrado, and Jeff Dean. 2013. Distributed representations of words and phrases and their compositionality. In *Advances in neural information processing systems*, pages 3111–3119.

Jeffrey Pennington, Richard Socher, and Christopher Manning. 2014. Glove: Global vectors for word representation. In *Proceedings of the 2014 conference on empirical methods in natural language processing (EMNLP)*, pages 1532–1543.

Pilar López Úbeda, Manuel Carlos Díaz Galiano, M Teresa Martín-Valdivia, and L Alfonso Urena Lopez. 2019. Using machine learning and deep learning methods to find mentions of adverse drug reactions in social media. In *Proceedings of the Fourth Social Media Mining for Health Applications (# SMM4H) Workshop & Shared Task*, pages 102–106.

VG Vinod Vydiswaran, Grace Ganzel, Bryan Romas, Deahan Yu, Amy Austin, Neha Bhomia, Socheatha Chan, Stephanie Hall, Van Le, Aaron Miller, et al. 2019. Towards text processing pipelines to identify adverse drug events-related tweets: University of michigan@ smm4h 2019 task 1. In *Proceedings of the Fourth Social Media Mining for Health Applications (# SMM4H) Workshop & Shared Task*, pages 107–109.

Sentence Transformers and Bayesian Optimization for Adverse Drug Effect Detection from Twitter

Oguzhan Gencoglu

Tampere University, Faculty of Medicine and Health Technology, Tampere, Finland

`oguzhan.gencoglu@tuni.fi`

Abstract

This paper describes our approach for detecting adverse drug effect mentions on Twitter as part of the Social Media Mining for Health Applications (SMM4H) 2020, Shared Task 2. Our approach utilizes multilingual sentence embeddings (sentence-BERT) for representing tweets and Bayesian hyperparameter optimization of sample weighting parameter for counterbalancing high class imbalance.

1 Introduction and Related Work

Automatic adverse drug reaction detection from social media has high significance to health informatics and pharmacovigilance due to fast, scalable, and diverse public health surveillance opportunities. Numerous studies have proposed natural language processing and machine learning solutions to detect mentions of adverse drug effects, especially from Twitter (Bian et al., 2012; Jiang and Zheng, 2013; Ginn et al., 2014; O'Connor et al., 2014; Katragadda et al., 2015; Egger et al., 2016; Korkontzelos et al., 2016; Rastegar-Mojarad et al., 2016; MacKinlay et al., 2017; Alimova and Tutubalina, 2017; Moh et al., 2017; Lee et al., 2017; Gupta et al., 2018a; Gupta et al., 2018b; Masino et al., 2018; Wu et al., 2019; Mesbah et al., 2019; Zhang et al., 2019; Alhuzali and Ananiadou, 2019). Proposed approaches in earlier studies vary from rule-based systems to deep learning. The challenge of detecting adverse drug effect mentions in Twitter remains unsolved due to lack of large-scale annotated datasets, rareness of relevant tweets among all tweets as well as ever-changing nature of the phenomenon.

The task consists of building separate adverse drug effect mention detection (binary classification) models for English, French, and Russian datasets from Twitter (9.25%, 1.61%, and 8.75% positive class prevalence, respectively). Detailed description of data and task can be found in (Klein et al., 2020).

2 Methods

2.1 Sentence Embeddings

We first preprocess the tweets by removing the usernames (in the form `@username`) and urls. We utilize recently introduced sentence-BERT (SBERT) models (Reimers and Gurevych, 2019) to represent each tweet as a *sentence embedding* instead of using standard BERT (Devlin et al., 2019) or its variants which work in a token-embedding manner. SBERT models are trained by a Siamese-network structure using BERT models to learn semantically meaningful sentence embeddings in an efficient manner. We utilize sentence-embedding versions of pretrained RoBERTa (Liu et al., 2019) model for English dataset (representation vectors of length 1024) and multilingual DistilBERT (Sanh et al., 2019) model (trained on 13 languages) for French and Russian datasets (representation vectors of length 512). We then train 3-layer (2 dense layers of size 256 and 32 with *ReLU* activation, respectively and an output layer with *sigmoid* activation) fully-connected neural networks using the sentence embeddings as input features. A Dropout rate of 0.5 is used between the dense layers. Model trainings are performed in a mini-batch manner for 100 epochs with a batch size of 32 with *Adam* optimizer (learning rate of 5×10^{-4}). Every

Proceedings of the 5ᵗʰ Social Media Mining for Health Applications (#SMM4H) Workshop & Shared Task, pages 161–164
Barcelona, Spain (Online), December 12, 2020.

run is performed in a 5-fold cross-validation manner and models at the epoch that maximizes the F_1 score on the validation split of the cross-validation is selected as final model.

2.2 Bayesian Optimization of Sample Weighting

As the problem at hand consists of highly imbalanced datasets, statistical balancing of positive and negative classes benefits the model training. Such balancing may be performed in various ways, e.g., with data augmentation. We chose to tackle the problem by increasing the contribution of the positive samples (corresponding to adverse effect mentions) to the loss function calculation. Each task requires a different weight coefficient (multiplier) for the positive samples, θ, as the prevalence of positive samples differ between datasets. We formulate the problem of finding the optimal multiplier, $\hat{\theta}$, as a Bayesian optimization problem:

$$\hat{\theta} = \operatorname*{argmax}_{\theta} f(\theta), \tag{1}$$

where $f(\theta)$ is the average of cross-validation F_1 scores, i.e., $\frac{1}{N}\sum_{i=1}^{N} F_1^i$. For our experiments $N = 5$ as we perform 5-fold cross-validation. We use a *Gaussian Process* for the surrogate model (Rasmussen, 2003) of the Bayesian optimization by which we emulate the statistical relationship between the positive sample weight coefficient and cross-validation performance, given a dataset.

3   Results

Results of our experiments and submissions can be examined from Table 1. Validation results correspond to local experiments, i.e., cross-validation with training data and inference on validation data. Test results (competition results) correspond to cross-validation with training + validation data and inference on test data from the ensembles of the cross-validation models (*mean* pooling). F_1 scores of 0.48, 0.17, and 0.42 are achieved for English, French, and Russian datasets, respectively.

| Task | SBERT model | $\hat{\theta}_{best}$ | Validation | | | Test | | | Comp. Avg. |
			P	**R**	F_1	**P**	**R**	F_1	F_1
English	RoBERTa	2.77	0.46	0.54	0.49	0.44	0.53	0.48	0.46
French	DistilBERT	19.67	0.19	0.25	0.22	0.15	0.20	0.17	0.07
Russian	DistilBERT	8.23	0.38	0.52	0.45	0.35	0.55	0.42	0.43

Table 1: Validation and competition test results of developed binary classification models for the 3 datasets (P = Precision, R = Recall).

4   Discussion and Conclusions

We chose to use pre-computed sentence embeddings instead of fine-tuning on token embeddings out of original BERT models due to its low computational requirements (Reimers and Gurevych, 2019). While it is possible to represent tweets with a standard BERT model as well (e.g. by averaging the token embeddings out of the output layer), such sentence representations were shown to be inferior to embeddings specifically trained for representing sentences (Reimers and Gurevych, 2019).

Bayesian optimization is beneficial in settings where the function to be minimized/maximized is a black-box function without a known closed-form, expensive to evaluate, and stochastic (Močkus, 1975). As $f(\theta)$ corresponds to cross-validation performance in our case, it is indeed a black-box function and computationally expensive to evaluate. Furthermore, it is a stochastic function as the same θ may give different outputs for different runs due to randomness in neural network weight initialization. That is our motive for employing Bayesian hyperparameter optimization for sample weighting. Advantage of using a Gaussian Process as the surrogate model is that it can provide uncertainty estimations. As expected, optimum positive sample weight parameters found by Bayesian optimization are different for each dataset and French dataset requires much higher balancing due to its extreme class inbalance. Our Bayesian optimization approach can be easily extended to optimization of other hyperparameters of the model including neural network architecture and other relevant design choices.

References

Hassan Alhuzali and Sophia Ananiadou. 2019. Improving classification of adverse drug reactions through using sentiment analysis and transfer learning. In *Proceedings of the 18th BioNLP Workshop and Shared Task*, pages 339–347.

Ilseyar Alimova and Elena Tutubalina. 2017. Automated detection of adverse drug reactions from social media posts with machine learning. In *International Conference on Analysis of Images, Social Networks and Texts*, pages 3–15. Springer.

Jiang Bian, Umit Topaloglu, and Fan Yu. 2012. Towards large-scale Twitter mining for drug-related adverse events. In *Proceedings of the 2012 International Workshop on Smart Health and Wellbeing*, pages 25–32.

Jacob Devlin, Ming-Wei Chang, Kenton Lee, and Kristina Toutanova. 2019. BERT: Pre-training of deep bidirectional transformers for language understanding. In *Proceedings of the 2019 Conference of the North American Chapter of the Association for Computational Linguistics: Human Language Technologies, Volume 1 (Long and Short Papers)*, pages 4171–4186.

Dominic Egger, Fatih Uzdilli, Mark Cieliebak, and L Derczynski. 2016. Adverse drug reaction detection using an adapted sentiment classifier. In *Proceedings of the Social Media Mining Shared Task Workshop at the Pacific Symposium on Biocomputing*.

Rachel Ginn, Pranoti Pimpalkhute, Azadeh Nikfarjam, Apurv Patki, Karen O'Connor, Abeed Sarker, Karen Smith, and Graciela Gonzalez. 2014. Mining Twitter for adverse drug reaction mentions: A corpus and classification benchmark. In *Proceedings of the Fourth Workshop on Building and Evaluating Resources for Health and Biomedical Text Processing*, pages 1–8. Citeseer.

Shashank Gupta, Manish Gupta, Vasudeva Varma, Sachin Pawar, Nitin Ramrakhiyani, and Girish Keshav Palshikar. 2018a. Co-training for extraction of adverse drug reaction mentions from tweets. In *European Conference on Information Retrieval*, pages 556–562. Springer.

Shashank Gupta, Sachin Pawar, Nitin Ramrakhiyani, Girish Keshav Palshikar, and Vasudeva Varma. 2018b. Semi-supervised recurrent neural network for adverse drug reaction mention extraction. *BMC Bioinformatics*, 19(8):212.

Keyuan Jiang and Yujing Zheng. 2013. Mining Twitter data for potential drug effects. In *International Conference on Advanced Data Mining and Applications*, pages 434–443. Springer.

Satya Katragadda, Harika Karnati, Murali Pusala, Vijay Raghavan, and Ryan Benton. 2015. Detecting adverse drug effects using link classification on Twitter data. In *IEEE International Conference on Bioinformatics and Biomedicine (BIBM)*, pages 675–679. IEEE.

Ari Z Klein, Ilseyar Alimova, Ivan Flores, Arjun Magge, Zulfat Miftahutdinov, Anne-Lyse Minard, Karen O'Connor, Abeed Sarker, Elena Tutubalina, Davy Weissenbacher, and Graciela Gonzalez-Hernandez. 2020. Overview of the fifth social media mining for health applications (#SMM4H) shared tasks at COLING 2020. In *Proceedings of the Fifth Social Media Mining for Health Applications (#SMM4H) Workshop & Shared Task*.

Ioannis Korkontzelos, Azadeh Nikfarjam, Matthew Shardlow, Abeed Sarker, Sophia Ananiadou, and Graciela H Gonzalez. 2016. Analysis of the effect of sentiment analysis on extracting adverse drug reactions from tweets and forum posts. *Journal of Biomedical Informatics*, 62:148–158.

Kathy Lee, Ashequl Qadir, Sadid A Hasan, Vivek Datla, Aaditya Prakash, Joey Liu, and Oladimeji Farri. 2017. Adverse drug event detection in tweets with semi-supervised convolutional neural networks. In *Proceedings of the 26th International Conference on World Wide Web*, pages 705–714.

Yinhan Liu, Myle Ott, Naman Goyal, Jingfei Du, Mandar Joshi, Danqi Chen, Omer Levy, Mike Lewis, Luke Zettlemoyer, and Veselin Stoyanov. 2019. RoBERTa: A robustly optimized BERT pretraining approach. *arXiv preprint arXiv:1907.11692*. version 1.

Andrew MacKinlay, Hafsah Aamer, and Antonio Jimeno Yepes. 2017. Detection of adverse drug reactions using medical named entities on Twitter. In *AMIA Annual Symposium Proceedings*, volume 2017, page 1215. American Medical Informatics Association.

Aaron J Masino, Daniel Forsyth, and Alexander G Fiks. 2018. Detecting adverse drug reactions on Twitter with convolutional neural networks and word embedding features. *Journal of Healthcare Informatics Research*, 2(1-2):25–43.

Sepideh Mesbah, Jie Yang, Robert-Jan Sips, Manuel Valle Torre, Christoph Lofi, Alessandro Bozzon, and Geert-Jan Houben. 2019. Training data augmentation for detecting adverse drug reactions in user-generated content. In *Proceedings of the 2019 Conference on Empirical Methods in Natural Language Processing and the 9th International Joint Conference on Natural Language Processing (EMNLP-IJCNLP)*, pages 2349–2359.

Jonas Močkus. 1975. On Bayesian methods for seeking the extremum. In *Optimization Techniques IFIP Technical Conference*, pages 400–404. Springer.

Melody Moh, Teng-Sheng Moh, Yang Peng, and Liang Wu. 2017. On adverse drug event extractions using Twitter sentiment analysis. *Network Modeling Analysis in Health Informatics and Bioinformatics*, 6(1):18.

Karen O'Connor, Pranoti Pimpalkhute, Azadeh Nikfarjam, Rachel Ginn, Karen L Smith, and Graciela Gonzalez. 2014. Pharmacovigilance on Twitter? mining tweets for adverse drug reactions. In *AMIA Annual Symposium Proceedings*, volume 2014, page 924. American Medical Informatics Association.

Carl Edward Rasmussen. 2003. Gaussian processes in machine learning. In *Summer School on Machine Learning*, pages 63–71. Springer.

Majid Rastegar-Mojarad, Ravikumar Komandur Elayavilli, Yue Yu, and Hongfang Liu. 2016. Detecting signals in noisy data-can ensemble classifiers help identify adverse drug reaction in tweets. In *Proceedings of the Social Media Mining Shared Task Workshop at the Pacific Symposium on Biocomputing*.

Nils Reimers and Iryna Gurevych. 2019. Sentence-BERT: Sentence embeddings using Siamese BERT-networks. In *Proceedings of the 2019 Conference on Empirical Methods in Natural Language Processing and the 9th International Joint Conference on Natural Language Processing (EMNLP-IJCNLP)*, pages 3982–3992, Hong Kong, China, November. Association for Computational Linguistics.

Victor Sanh, Lysandre Debut, Julien Chaumond, and Thomas Wolf. 2019. DistilBERT, a distilled version of BERT: smaller, faster, cheaper and lighter. *arXiv preprint arXiv:1910.01108*. version 4.

Chuhan Wu, Fangzhao Wu, Zhigang Yuan, Junxin Liu, Yongfeng Huang, and Xing Xie. 2019. MSA: Jointly detecting drug name and adverse drug reaction mentioning tweets with multi-head self-attention. In *Proceedings of the Twelfth ACM International Conference on Web Search and Data Mining*, pages 33–41.

Tongxuan Zhang, Hongfei Lin, Yuqi Ren, Liang Yang, Bo Xu, Zhihao Yang, Jian Wang, and Yijia Zhang. 2019. Adverse drug reaction detection via a multihop self-attention mechanism. *BMC Bioinformatics*, 20(1):479.

Sentence Contextual Encoder with BERT and BiLSTM for Automatic Classification with Imbalanced Medication Tweets

Olanrewaju Tahir Aduragba, Jialin Yu, Gautham Senthilnathan and **Alexandra Cristea**
Department of Computer Science
Durham University, Durham, UK
{olanrewaju.m.aduragba, jialin.yu}@durham.ac.uk
{gautham.senthilnathan, alexandra.cristea}@durham.ac.uk

Abstract

This paper details the system description and approach used by our team for the SMM4H 2020 competition, Task 1. Task 1 targets the automatic classification of tweets that mention medication. We adapted the standard BERT pretrain-then-fine-tune approach to include an intermediate training stage with a biLSTM architecture neural network acting as a further fine-tuning stage. We were inspired by the effectiveness of within-task further pre-training and sentence encoders. We show that this approach works well for a highly imbalanced dataset. In this case, the positive class is only 0.2% of the entire dataset. Our model performed better in both F1 and precision scores compared to the mean score for all participants in the competition and had a competitive recall score.

1 Introduction

Social media is ubiquitous, a continuous part of our daily lives; it offers new ways of communication and contains important data for various disciplines (Pershad et al., 2018). This is especially crucial for health related sectors. Twitter posts are now recognised as an important source of patient-generated data, providing unique insights into population health (Nguyen et al., 2017). However, tweets contain a lot of noise and are imbalanced in their nature. A fundamental step towards incorporating Twitter data in pharmacoepidemiologic research is to automatically recognise medication mentions in tweets. Hence Task 1 in SMM4H 2020 (Ari Z. Klein and Gonzalez-Hernandez, 2020) is proposed based on a similar task from 2018 (Weissenbacher et al., 2018) to address these challenges. The task is aimed at identifying if tweets contain information about medications using a highly skewed, imbalanced training dataset in which only 0.2% contains a positive label.

2 Methodology

2.1 Data

The *training data* provided by SMM4H 2020 consists of 69,272 training data, out of which only 181 are positive tweets, and 69,091 are negative tweets. I.e., it contains binary labelled unique tweets that mention a medication or dietary supplement (annotated as "1") and tweets that do not (annotated as "0"). Compared with the same task in 2018, where artificially balanced tweet data was provided, the current training data consist of highly imbalanced tweets, in which only 0.2% mentioned a medication (annotated as "1"), corresponding to a more realistic real-world scenario. The *test data* provided contains 29,687 examples. Additionally, the previously mentioned artificially balanced tweet dataset used for the SMM4H 2018 Task 1 (Weissenbacher et al., 2018), was released to participants, containing 2,367 unique tweets with 1,212 positive and 1,155 negative labels.

2.2 Data Cleaning and Pre-processing

As the distinct way the reference to medication is made is not important in this binary classification tasks, we applied various standard pre-processing and cleaning methods on the tweets, as follows. Be-

Proceedings of the 5th Social Media Mining for Health Applications (#SMM4H) Workshop & Shared Task, pages 165–167
Barcelona, Spain (Online), December 12, 2020.

fore training, we processed the datasets using the Ekphrasis library (Baziotis et al., 2017). The library provides text preprocessing operations optimised for social media texts. Each tweet was normalised, by replacing all Twitter usernames and URLs with common tokens, such as <user>and <url>. Hashtags, emoticons, elongated and repeated words were further annotated with special tokens using the 'Social Tokenizer' provided in the above library. As an extra preprocessing step, we used spell correction and word segmentation tools from the same library to correct misspelled words and split long texts into more meaningful sub word forms.

2.3   Preliminary Fine-tuning of the Model on a Balanced Task

Before training with the SMM4H 2020 Task 1 training data, we use the artificially balanced data from 2018 to fine-tune our model; here, we apply BERT (Devlin et al., 2018), as the most advanced training network available currently, with a learning rate of 2e-5. As BERT is not a classifier, following the fine-tuning process (Devlin et al., 2018; Sun et al., 2019), we use the last layer of BERT and build a linear layer on top of it, as a classification task. We perform this preliminary fine-tuning as we expect that the balanced dataset allows the BERT model to better understand and capture the generality of the data distribution for tweets that mention medications.

2.4   Target Task Training with BERT Feature Information

After the preliminary training of BERT, the model can be considered as containing the contextual information of medication tweets data. Inspired by the effectiveness of the within-task further pre-training and sentence encoder (Phang et al., 2018; Sun et al., 2019), we further experimented with the idea of a supplementary training phase on a data-rich supervised task. This process involves training BERT using the output as features in our target task. We experimented with different outputs from the last hidden states. As recommended in the original BERT paper (Devlin et al., 2018), we experimented with the feature-based approach using the activations from the last hidden layer, the sum of all 12 hidden layers, the sum of the last four hidden layers and the concatenation of the last four hidden layers as input into our classification model.

The model parameters learned from the intermediate task served as contextualised sentence embeddings (Phang et al., 2018) to initialise our target task model. We feed these embedding to a bidirectional LSTM (biLSTM) model (Graves and Schmidhuber, 2005) to classify the tweets. We experimented with the output of each encoder layer, to see which vector provides the best contextualised embedding. These experiments showed that the sum of the final four hidden layers of the pre-trained model provides the best embedding. We then trained the biLSTM model for 5 epochs to reach optimal performance using a binary cross-entropy loss.

For the intermediate training, we utilise the features generated by BERT as the contextual sentence embedding, similar to word embedding. We compared the difference in performance by freezing and unfreezing the feature vectors from BERT. This comparison suggest that unfreezing achieves better performance. This can be considered as a transfer learning phase which results in a boost of performance.

3   Results

As mentioned previously, the information from BERT can be considered as a sentence embedding to the supplementary BiLSTM model; we have the option to either freeze or unfreeze the BERT model parameters in the target task training. Freezing the parameters means that we use a fixed prior knowledge from the balanced dataset as a sentence embedding to the BiLSTM model, while unfreezing the parameters allows the parameters in the sentence embedding to update with the new data, similarly to how the NLP field uses pre-trained word embeddings and then fine-tunes them with new data. The results from the experiments, when we freeze and unfreeze the BERT part, are presented in Table 1 and Table 2, respectively. All the results in both tables are calculated based on the unbalanced data. As said, the best performing model on the imbalanced dataset results from concatenating the final four layers. Our system achieved an F1 score on the positive (minority) class of 0.7164, with 0.8421 precision and 0.6234 recall on the test dataset. The mean score for this task was an F1 score on the positive class of 0.6646, with

0.7007 precision and 0.7039 recall.

4   Conclusion

The results of our experiments suggest that extracting features from language models and using them for intermediate modelling can provide a good platform to study the automatic classification of medical tweets, even if the final target dataset is highly imbalanced. Moreover, interestingly, various layers of features from language models can be used as a sequence in the representation, and this further confirms that different layers of representation in complex language models, such as BERT, contain various levels of information semantics.

Layers	F1	Precision	Recall	TP/FP/FN
concatenate last 4 hidden layers	0.78	0.79	0.77	27/7/8
last hidden layer*	**0.81**	**0.84**	**0.77**	**27/5/8**
sum of last 4 hidden layers	0.75	0.78	0.71	25/7/10
second to last hidden layers	0.81	0.84	0.77	27/5/8

Table 1: Experiments results for freezing parameters of BERT

Layers	F1	Precision	Recall	TP/FP/FN
concatenate last 4 hidden layers*	**0.88**	**0.97**	0.80	28/1/7
last hidden layer	0.83	0.81	**0.86**	**30/7/5**
sum of last 4 hidden layers	0.85	0.83	**0.86**	30/6/5
second to last hidden layer	0.84	0.85	0.83	29/5/6

Table 2: Experiments results for not freezing parameters of BERT

References

Ivan Flores Arjun Magge Zulfat Miftahutdinov Anne-Lyse Minard Karen O'Connor Abeed Sarker Elena Tutubalina Davy Weissenbacher Ari Z. Klein, Ilseyar Alimova and Graciela Gonzalez-Hernandez.   2020. Overview of the fifth social media mining for health applications (smm4h) shared tasks at coling 2020. In *Proceedings of the Fifth Social Media Mining for Health Applications (#SMM4H) Workshop & Shared Task.*

Christos Baziotis, Nikos Pelekis, and Christos Doulkeridis. 2017. Datastories at semeval-2017 task 4: Deep lstm with attention for message-level and topic-based sentiment analysis. In *Proceedings of the 11th International Workshop on Semantic Evaluation (SemEval-2017)*, pages 747–754, Vancouver, Canada, August. Association for Computational Linguistics.

Jacob Devlin, Ming-Wei Chang, Kenton Lee, and Kristina Toutanova. 2018. Bert: Pre-training of deep bidirectional transformers for language understanding. *arXiv preprint arXiv:1810.04805.*

Alex Graves and Jürgen Schmidhuber. 2005. Framewise phoneme classification with bidirectional lstm and other neural network architectures. *Neural networks*, 18(5-6):602–610.

Quynh C Nguyen, Kimberly D Brunisholz, Weijun Yu, Matt McCullough, Heidi A Hanson, Michelle L Litchman, Feifei Li, Yuan Wan, James A VanDerslice, Ming Wen, et al. 2017. Twitter-derived neighborhood characteristics associated with obesity and diabetes. *Scientific reports*, 7(1):1–10.

Yash Pershad, Patrick Hangge, Hassan Albadawi, and Rahmi Oklu. 2018. Social medicine: Twitter in healthcare. *Journal of Clinical Medicine*, 7(6):121, May.

Jason Phang, Thibault Févry, and Samuel R Bowman. 2018. Sentence encoders on stilts: Supplementary training on intermediate labeled-data tasks. *arXiv preprint arXiv:1811.01088.*

Chi Sun, Xipeng Qiu, Yige Xu, and Xuanjing Huang. 2019. How to fine-tune bert for text classification? In *China National Conference on Chinese Computational Linguistics*, pages 194–206. Springer.

Davy Weissenbacher, Abeed Sarker, Michael Paul, and Graciela Gonzalez. 2018. Overview of the third social media mining for health (smm4h) shared tasks at emnlp 2018. In *Proceedings of the 2018 EMNLP Workshop SMM4H: The 3rd Social Media Mining for Health Applications Workshop & Shared Task*, pages 13–16.

CLaC at SMM4H 2020: Birth defect mention detection

Parsa Bagherzadeh, Sabine Bergler
CLaC Labs, Concordia University
Montreal, Canada
{p_bagher, bergler} @cse.concordia.ca

Abstract

For the detection of personal tweets, where a parent speaks of a child's birth defect, CLaC combines ELMo word embeddings and gazetteer lists from external resources with a GCNN (for encoding dependencies), in a multi layer, transformer inspired architecture. To address the task, we compile several gazetteer lists from resources such as MeSH and GI. The proposed system obtains .69 for μF1 score in the SMM4H 2020 Task 5 where the competition average is .65.

Introduction Tweets potentially offer a record that is distilled from messages that are written for other purposes. They contain many particular, directly observed experiences, which are of great interest to epidemiologists and health monitors. Situating congenital abnormalities (birth defects) in time and space is important to identify causes of birth defects, find opportunities to prevent them, and improve the health of those living with them. Therefore, identifying tweets mentioning birth defect experiences by a family is helpful, which motivated the SMM4H 2020 Task 5 (Klein et al., 2020).

This paper reports our effort toward building a predictive system to address the task. We try to leverage medical annotations, family relation annotations, as well as dependency relations in our model.

Task description and data The SMM4H 2020 task 5 is a 3-way classification problem. Class 1 tweets refer to the user's child and indicate that he/she has a birth defect. Class 2 tweets are ambiguous about whether someone is the user's child or has a birth defect mentioned in the tweet. Class 3 tweets merely mention birth defects. Training and development sets provided by the organizers include 14705 and 3677 tweets respectively and the evaluation set includes 4603 tweets.

Preprocessing and annotations Tweets are tokenized using the ANNIE tweet tokenizer (Cunningham et al., 2002) as well as the hashtag tokenizer (Maynard and Greenwood, 2014). URLs and user mentions are removed. We use the Stanford parser (Klein and Manning, 2003) to encode dependencies and ANNIE *Names* for named entity recognition.

We compiled gazetteer lists for birth defect detection from MeSH (Lipscomb, 2000), namely *BirthDef*, a list of congenital, hereditary, and neonatal diseases and abnormalities compiled from MeSH C16 and *PregComp*, a list of pregnancy complications compiled from MeSH C13.703. Each MeSH sub-tree is traversed depth-first and all entry terms for each heading (both for internal nodes and leaves) are added to the gazetteers. A word list of family terms, *FamilyRel*, that includes terms like *son, daughter, cousin, etc.* is also complied. Moreover, a list *Acquaintance* which contains terms like *friend, colleague, neighbor, etc.* is extracted from the General Inquirer (Stone et al., 1966).

These four gazetteer lists form the parameter *Annotation* or $A = \{BirthDef, PregComp, FamilyRel, Acquaintance\}$ and form the four annotation types.

Deep model Our system consists of four stacked layers. Each layer $l \geq 2$ receives token representasions h_i^{l-1} and passes new represensations h_i^l to the next layer. The layers are:

Layer 1: annotation embedding and token embedding. Annotation types are embedded using a matrix $M \in \mathbb{R}^{4 \times 1024}$ (four rows corresponding to four annotation types). M is initialized randomly

Proceedings of the 5th Social Media Mining for Health Applications (#SMM4H) Workshop & Shared Task, pages 168–170
Barcelona, Spain (Online), December 12, 2020.

and learned as a parameter during training for the main classification task. To embed tokens we use ELMo (Peters et al., 2018). The new representations h_i^1 is obtained by summing annotation embeddings to token embeddings (see Figure 1).

Layer 2: the encoder part of the Transformer (Vaswani et al., 2017). The encoder gets the representations h_i^1 and outputs representations h_i^2. The number of heads in the multi-head attention is $n_{heads} = 4$ and the dimensionality of the feed-forward layer is $d_{FF} = 1024$.

Layer 3: graph convolutional network (GCNN) (Kipf and Welling, 2017) for dependency relation encoding following (Marcheggiani and Titov, 2017). In GCCN each token is represented based on its adjacent tokens in dependency parse by $h_i^l = ReLU(\sum_{j \in \mathcal{N}(i)} W_{L(i,j)} h_j^l + b)$ where $\mathcal{N}(i)$ is the set of tokens adjacent to token i and $L(j, i)$ is the label of the arc from token j to token i. Note that the network is not tied, i.e. $W_{L(i,j)}$ depends on the arc labels. GCNN receives h_i^2 and outputs token-wise representations h_i^3.

Layer 4: attention (Bahdanau et al., 2015), which calculates importance scores $e_i = w_{att}^T h_i^3$ using a latent context vector w_{att} and normalizes the scores using softmax ($\alpha_i = \frac{exp(e_i)}{\sum_j e_j}$) for a weighted sum $H = \sum_i \alpha_i * h_i^3$.

Attention produces an output vector $H \in \mathbb{R}^{1024}$ for the tweet, which is fed into three decision neurons for classifying into the three output classes.

Our model is implemented using PyTorch (Paszke et al., 2017) and optimized with the Adam optimizer (Kingma and Ba, 2015) with a learning rate of $lr = 0.0005$ for 7 to 10 epochs.

	h_1^1	h_2^1	h_3^1	h_4^1	h_5^1	h_6^1	h_7^1	h_8^1	h_9^1
	=	=	=	=	=	=	=	=	=
Annotation Embedding	M_{Name}	0	$M_{BirthDef}$	0	0	0	0	0	0
	+	+	+	+	+	+	+	+	+
Token Embedding	E_{Milo}	E_{has}	$E_{Hydrocephalus}$	$E_{causing}$	E_{him}	E_{to}	E_{need}	E_{brain}	E_{shunt}
Input	Milo	has	Hydrocephalus	causing	him	to	need	brain	shunt

Figure 1: Additive annotation embedding. M_a is the row in M that corresponds to annotation type $a \in A$

Submitted system   Due to a submission issue, only one of three planned configurations was submitted.

Ablation studies on our validation data showed that all four layers of the architecture add to the overall performance:

Table 1: Ablations and performance. TE denotes token embedding, AE denotes annotation embeddings

System	Layer	Validation set					Test set		
		Class 1 F1	Class 2 F1	μP	μR	μF1	μP	μR	μF1
TE+Trans	1,2,4	.70	.63	.72	.60	.65	-	-	-
TE+GCNN	1,3,4	.72	.64	.74	.60	.67	-	-	-
TE+AE+Trans	1,2,4	.75	.65	.70	.69	.70	-	-	-
TE+AE+Trans+GCNN	1,2,3,4	.76	.68	.72	.69	.71	.71	.67	.69
competition mean	-	-	-	-	-	-	.62	.68	.65

The results on our development set suggest that the best performance is achieved by the full system (indicated in boldface). The official competition results are provided on the right side of Table1. The micro average scores for our submission is close to its development score, which we consider an important sign of robustness in our system.

Conclusion   For the SMM4H 2020 Task 5 competition we proposed a multi-layer system to leverage gazetteer list annotations as well as a dependency parse. The reported experiments here suggest that incorporating external knowledge in form of textual annotation has the potential to enhance the performance of models trained on moderate-sized training sets in a robust and efficient manner.

References

Dzmitry Bahdanau, Kyunghyun Cho, and Yoshua Bengio. 2015. Neural machine translation by jointly learning to align and translate. In *Proceedings of the 3rd International Conference on Learning Representations, ICLR'15*.

H Cunningham, D Maynard, K Bontcheva, and V Tablan. 2002. GATE: A framework and graphical development environment for robust NLP tools and applications. In *ACL'02*.

DP Kingma and J Ba. 2015. Adam: A method for stochastic optimization. In *3Proceedings of the 3rd International Conference for Learning Representations (ICLR 2015)*. arXiv:1412.6980 [cs.LG].

Thomas N Kipf and Max Welling. 2017. Semi-supervised classification with graph convolutional networks. In *In Proceedings of ICLR*.

Dan Klein and Christopher D. Manning. 2003. Accurate unlexicalized parsing. In *Proceedings of the 41st Meeting of the Association for Computational Linguistics*.

Ari Z. Klein, Ivan Flores, Arjun Magge, Anne-Lyse Minard, Karen O'Connor, Abeed Sarker, Elena Tutubalina, Davy Weissenbacher, and Graciela Gonzalez-Hernandez. 2020. Overview of the fifth Social Media Mining for Health Applications (SMM4H) Shared Tasks at COLING 2020. In *Proceedings of the Fifth Social Media Mining for Health Applications (SMM4H) Workshop & Shared Task*.

Carolyn E Lipscomb. 2000. Medical subject headings (MeSH). *Bulletin of the Medical Library Association*, 88(3).

Diego Marcheggiani and Ivan Titov. 2017. Encoding sentences with graph convolutional networks for semantic role labeling. In *Proceedings of the 2017 Conference on Empirical Methods in Natural Language Processing*, pages 1506–1515.

DG Maynard and MA Greenwood. 2014. Who cares about sarcastic tweets? Investigating the impact of sarcasm on sentiment analysis. In *LREC 2014*.

Adam Paszke, Sam Gross, Soumith Chintala, Gregory Chanan, Edward Yang, Zachary DeVito, Zeming Lin, Alban Desmaison, Luca Antiga, and Adam Lerer. 2017. Automatic differentiation in PyTorch. In *NIPS 2017*.

Matthew E Peters, Mark Neumann, Mohit Iyyer, Matt Gardner, Christopher Clark, Kenton Lee, and Luke Zettlemoyer. 2018. Deep contextualized word representations. In *Proceedings of NAACL-HLT*, pages 2227–2237.

Philip J Stone, Dexter C Dunphy, and Marshall S Smith. 1966. *The general inquirer: A computer approach to content analysis*. MIT press.

Ashish Vaswani, Noam Shazeer, Niki Parmar, Jakob Uszkoreit, Llion Jones, Aidan N Gomez, Łukasz Kaiser, and Illia Polosukhin. 2017. Attention is all you need. In *Advances in neural information processing systems*.

Association for Computational Linguistics
209 N. Eighth Street
Stroudsburg, Pennsylvania 18360

ISBN 978-1-7138-2820-4